Nonlinear diffusive waves

Nonlinear diffusive waves

P. L. Sachdev

Professor, Department of Applied Mathematics,
Indian Institute of Science, Bangalore

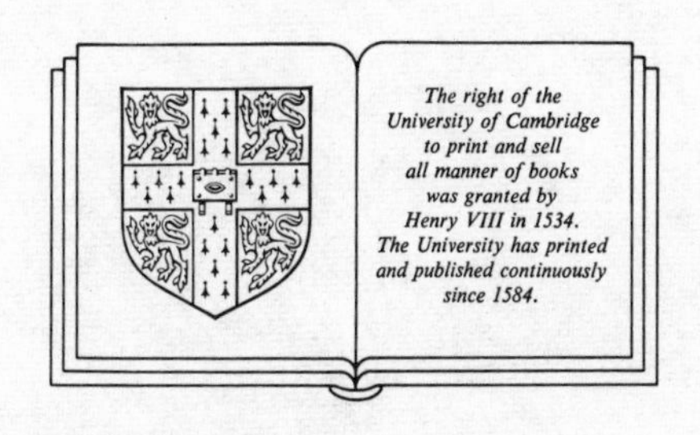

CAMBRIDGE UNIVERSITY PRESS

Cambridge

New York New Rochelle Melbourne Sydney

Published by the Press Syndicate of the University of Cambridge
The Pitt Building, Trumpington Street, Cambridge CB2 1RP
32 East 57th Street, New York, NY 10022, USA
10 Stamford Road, Oakleigh, Melbourne 3166, Australia

First published 1987

Printed in Great Britain at the University Press, Cambridge

British Library cataloguing in publication data

Sachdev, P. L.
Nonlinear diffusive waves.
1. Wave-motion, Theory of 2. Burgers
equation 3. Diffusion
I. Title
531'.1133 QA927

Library of Congress cataloguing in publication data

Sachdev, P. L.
Nonlinear diffusive waves.
Bibliography
Includes index.
1. Nonlinear waves. I. Title.
QA927.S214 1987 531'.1133 86–18397

ISBN 0 521 26593 2

TM

Contents

Preface

I was fortunate to receive help from many. Professor Sir James Lighthill FRS provided the impetus to undertake this venture. Dr Allen Tayler helped my rather amorphous ideas assume a precise form. Professor A. Richard Seebass, many years ago, introduced me to the mysteries of Burgers' equation and its kindred class. I had the benefit of very fruitful discussions with Professors D. G. Crighton and J. D. Murray. Professor Colin Rogers read through parts of the manuscript. Professors P. N. Kaloni and M. C. Singh were excellent hosts during my sojourns at the Universities of Windsor and Calgary, where I wrote part of the book. My student, Mr K. R. C. Nair, carried out much of the computations reported in the final chapter and assisted in various other ways. Ms Thelma Stanley, with good cheer and patience, typed the manuscript and many a change I made in it. Finally, I owe a lot to my wife Rita, who cared and comforted.

I gratefully acknowledge the financial assistance provided for the preparation of the manuscript by the Curriculum Development Cell, established at the Indian Institute of Science by the Ministry of Education and Culture, Government of India.

In conclusion, I thank Dr David Tranah of Cambridge University Press, for his courtesy and thoughtful consideration during the entire course of the publication of this monograph.

1 Introduction and overview

Waves are ubiquitous in nature. They have been studied in the last couple of decades in such diverse forms and varied fields that they may now be said to constitute a new discipline – the science of waves (Lighthill, 1978). This wide and varied interest in waves has been particularly helped by the appearance of that strange entity, the soliton. The wave adopts such diverse forms that it is difficult to present a precise unifying definition. However, we may agree that waves (or disturbances), in an otherwise quiet or uniformly moving medium, have propagation properties and therefore involve the variable time, and have distinct features such as crests and troughs which themselves move with definite speeds. It should, however, be noted that not all waves are oscillatory. Thus, shock waves and solitary waves are not oscillatory. Nevertheless, these are regarded as (nonlinear) entities of great physical importance.

Two major types of waves have been distinguished (Whitham, 1974). The first is called hyperbolic and requires the system of n governing partial differential equations to have n real characteristic directions and correspondingly n linearly independent left eigenvectors of the relevant matrix (Courant & Hilbert, 1962). The second type of waves, called dispersive, are categorised by a real dispersion relation connecting the frequency and wave number (Bhatnagar, 1979). These definitions are broadened suitably to apply to partial differential equations with variable coefficients as well as nonlinear ones.

There is another type of nonlinear wave which is diffusive and which is epitomised by the equation

$$u_t + uu_x = \frac{\delta}{2} u_{xx}. \tag{1.1}$$

This is the celebrated Burgers equation. Here δ is a (small) coefficient of viscous diffusion. This is a nonlinear parabolic equation and describes in a simple manner a balance between nonlinear convection and linear diffusion or dissipation. This equation and its generalisations – scalar as well as vector – describe phenomena in such a variety of situations that they deserve a distinct categorisation, namely nonlinear diffusive wave equations. It must be recognised that, in the limit of $\delta \to 0$, eq. (1.1) goes into a scalar hyperbolic equation. Indeed, eq. (1.1) was suggested as a model to describe the structure of shock waves in gas dynamics, which is missed by the hyperbolic type of equations for which the shock appears as a sharp discontinuity.

A considerable part of the present monograph is devoted to the discussion of the Burgers equation and its generalisations (GBEs). However, it must be emphasised that the Burgers equation by no means exhausts the nonlinear diffusive phenomena. To stress this point and to bring out the contrast between nonlinear convective diffusive equations and those without convection, we discuss in detail several other nonlinear model equations in chapter 4. These include Fisher's equation, a nonlinear heat equation etc. For most of this monograph, we consider scalar equations only.

Chapter 2 begins with a heuristic derivation of the Burgers equation. This is followed by an order-of-magnitude analysis of the Navier–Stokes equations to derive a coupled system of two equations describing one-dimensional waves of finite amplitude in a viscous and heat-conducting gas, a generalised Burgers system. This system, under further approximation, delivers the Burgers equation. After a brief historical account of this equation, and the Hopf–Cole transformation which exactly linearises it to the heat equation, a pure initial value problem is posed and solved in a simple manner via the corresponding problem for the heat equation. Special solutions describing important physical situations such as the travelling shock wave, the single hump, the N wave and the periodic profile are derived. Continuous or distribution functions as initial conditions are assumed, and physical (or dimensional) arguments pointing to the similarity form of the solutions for travelling shock and single hump are described. These special solutions are carefully analysed in each of the several temporal and spatial domains that arise from considerations of the importance and balance of different terms. This is motivated by the desire to find analytic solutions, at least in some of the domains, of GBEs in subsequent chapters, for which no Hopf–Cole-like transformation exists. Although most of the earlier investigations relate to initial value problems, we also pose, in the semi-infinite domain, a boundary value problem for the Burgers equation and use a certain equivalence theorem for the heat equation, between initial value problem over the whole real line and a boundary value problem over the positive real line, to recover earlier solutions now arising from certain boundary conditions.

The Burgers equation is very important from the mathematical point of view as a canonical form since it highlights clearly the nature of analytic solutions in various temporal and spatial domains, which become available due to the Hopf–Cole transformation. The equations that arise in physical applications are more general than the Burgers equation and do not, in general, admit exact analytic solutions. Chapter 3 treats GBEs. After a brief review of the singular perturbation methods, we employ them to find a uniformly valid solution to order δ for the GBE, which, besides usual terms, has a linear damping as an additional term. We then derive from the Navier–Stokes equations a model which combines the effect of spherical or cylindrical expansion besides nonlinearity, viscous diffusion and heat conduction. This is achieved by using the method of multiple scaling. The nonplanar GBEs are studied analytically, as far as possible, using matched asymptotic expansions, in certain of the temporal domains for the sharp N wave initial profile. Reference is made to the gaps which still remain unbridged. A kindred discussion relates to the solution of the harmonic boundary value problem arising from a piston motion. For this purpose, it makes more sense to pose a boundary value problem in a semi-infinite domain, altering in the process the basic equation so that the roles of distance and (retarded) time are interchanged. Significant physical and mathematical consequences of the solution of the harmonic problem are carefully analysed. Here, even though we treat the standard Burgers equation, we include it in the chapter on generalised Burgers equations because of the complexity introduced by the boundary conditions. Generalised Hopf–Cole transformations are used to find a whole class of GBEs which may be changed into linear parabolic equations with variable coefficients. Some of the physically relevant equations are identified and their special solutions are discussed. Prominent among these is an inhomogeneous Burgers equation which occurs in several physical contexts.

While a major part of this monograph is concerned with equations of Burgers type which have a convective term as an important element responsible for wave steepening and shock phenomena, chapter 4 deals with a few representative nonlinear diffusion equations wherein the convective term is absent. This has been done for two reasons: firstly, to visualise how other nonlinear diffusion phenomena compare with that simulated by the equations of Burgers type, and, secondly, to prepare the ground for the discussion of stability (or intermediate asymptotic) analysis for a variety of nonlinear diffusion equations. The equations in this chapter include Fisher's equation, a nonlinear heat equation and an equation from plasma physics which has a (spatially) variable coefficient besides nonlinearity. While for the former equations the special solutions we study belong to the usual similarity form, the latter have a product form. Indeed, the reader will notice a strong undercurrent of the similarity approach in the entire course of the present monograph. This is partly due to the author's bias and partly due to the conviction that the similarity analysis leads to some of the most bona fide exact solutions of nonlinear problems. Historically, the similarity/product solutions were viewed as special solutions of nonlinear partial differential equations, their chief distinction being that they were governed by nonlinear ordinary differential equations which could be solved more conveniently either in a closed form or numerically. This viewpoint has since undergone a change. These special solutions represent what are referred to as intermediate asymptotics, 'describing the behaviour of the solutions to the original equations for a wide class of initial, boundary and mixed problems, away from the boundaries of the region of independent variables or, alternatively, in a region where in a sense the solution is no longer dependent on the details of the initial and/or boundary conditions but is still far from being in a state of equilibrium' (Barenblatt and Zel′dovich, 1972). In chapter 4, we discuss the role of the similarity/product type of solutions as intermediate asymptotics for a few representative nonlinear diffusion equations. That is to say we study the solutions of a class of initial and/or boundary value problems for these equations which evolve into self-similar solutions, as well as the manner and mode of such evolution. The equations that we treat here include a nonlinear heat equation, a nonlinear diffusion equation of plasma physics with a variable coefficient and the GBEs in spherical and cylindrical symmetry.

The analytical studies reported in chapter 3 clearly bring out the gaps in the understanding of the solutions of the GBEs. For example, for the non-planar GBE, there are several domains – the embryonic shock region and the infinitely long (in time) one beyond the Taylor-shock region – for which the analytical form of the solution seems difficult to obtain. Indeed, even the final phase of the N wave propagation, which is essentially linear, remains undetermined to the extent of an unknown multiplication factor for the cylindrically symmetric case. Therefore, there is a need to have a thorough understanding of 'good' numerical solutions of these equations which might, in turn, suggest the analytic form of the solution. In the final chapter of this monograph, we discuss two numerical techniques – implicit finite difference and the pseudo-spectral (accurate space differencing) – for three nonlinear diffusion equations, namely Fisher's equation, the GBEs in spherical and cylindrical symmetries and the GBE with a damping term. The need for using the pseudo-spectral approach becomes imperative for discontinuous initial data which the implicit scheme is not able to handle in an effective and accurate manner and which the pseudo-spectral scheme is. However, once the discontinuous profile has smoothed out and has settled down, say, to one with a Taylor shock, the implicit difference scheme can take over and deliver accurate results with great economy, in comparison with the pseudo-spectral approach which, though very accurate, is expensive in terms of computer time. The numerical techniques help understand the intermediate asymptotic nature of the travelling wave solution of Fisher's equation, the decay of spherical and cylindrical N waves and of the single hump initial profile evolving under a GBE with a damping term or under non-planar GBEs.

The present monograph is almost entirely devoted to scalar diffusive equations. Reference may be made to Smoller (1983) for systems of equations describing, in particular, reaction-diffusion processes. Moreover, we have restricted ourselves mostly to the gas dynamic context of the Burgers equation and its generalisations. For applications to turbulent flows, we refer the reader to Burgers (1974), Gurbatov *et al.* (1983) and Qian (1984). The review article of Gurbatov *et al.* contains a large bibliography.

While the major applications of the nonlinear diffusive equations have been drawn from gas dynamics, it will become apparent from the references that they occur frequently in many other areas such as plasma physics, heat conduction, elasticity, biomathematics etc. Therefore, the material in this monograph should be useful to scientists and engineers working in these areas. The treatment of the problems in the monograph is mainly applied mathematical in nature; however, the physical explanation is also briefly provided.

The prerequisites for the present monograph are a basic course in gas dynamics, and a fair knowledge of ordinary and partial differential equations. In particular, familiarity with the theory of parabolic partial differential equations will be found helpful.

2 The Burgers equation

2.1 Introduction

Wave phenomena are, in general, governed by nonlinear systems of partial differential equations subject to certain physically motivated initial and/or boundary conditions. The Navier–Stokes equations represent a typical example of such a system. These systems, in most cases, cannot be solved by exact analytic approaches. Indeed, even the numerical solution of these systems poses severe difficulties. Thus, in recent years, there have been attempts to derive simpler equations using perturbation methods, which retain from the larger systems the essentials of the physical problems and which hold over extensive spatial and temporal domains. The fact that these model equations commonly appear in a variety of physical contexts attests to their importance. Furthermore, recent investigations have shown that various model equations governing similar physical phenomena enjoy unifying mathematical properties. The best known examples are the Burgers equation

$$u_t + uu_x = \frac{\delta}{2}u_{xx}, \tag{2.1}$$

and the Korteweg–deVries equation

$$u_t + \sigma uu_x + u_{xxx} = 0. \tag{2.2}$$

While our main concern will be with eq. (2.1), we shall often compare and contrast eqs. (2.1) and (2.2) and their solutions, since the study of these *apparently* similar equations has provided mutual enrichment and led to important results for kindred classes of equations.

We commence our discussion with the system describing plane compressible flows in an ideal (polytropic) gas ignoring dissipative effects, viz.

$$\rho_t + v\rho_x + \rho v_x = 0, \tag{2.3}$$

$$\rho(v_t + vv_x) + p_x = 0, \tag{2.4}$$

$$p = k\rho^\gamma, \quad S = \text{constant}. \tag{2.5}$$

Here ρ, v and p are the density, particle velocity and pressure, respectively, depending on the spatial co-ordinate x and time t. S is entropy, assumed to be constant, $\gamma = c_p/c_v$, the ratio of specific heats, and k is a constant. If we consider a small disturbance over a uniform quiescent medium ($u = 0$, $\rho = \rho_0, p = p_0$), we may linearise the system (2.3)–(2.5) and obtain an equation, by suitable elimination, describing the disturbances in any of the variables

$$v' = v, \quad p' = p - p_0, \quad \rho' = \rho - \rho_0: \tag{2.6}$$

$$\rho'_{tt} - a_0^2 \rho'_{xx} = 0, \tag{2.7}$$

$$a_0^2 = a^2(\rho_0, S_0), \quad a \text{ being the speed of sound}$$

(see Courant and Friedrichs (1948, p. 19)). Eq. (2.7) is the 'standard' wave equation which has the general solution

$$\rho' = \phi(x - a_0 t) + \psi(x + a_0 t) \tag{2.8}$$

where ϕ and ψ are arbitrary appropriately differentiable functions of their arguments. The relation (2.8) describes the solution ρ of any initial value problem for eq. (2.7), the functions ϕ and ψ being determined by the initial conditions. These functions describe waves splitting from the initial conditions and moving to the right and the left, respectively, with speed a_0. We shall subsequently refer to solutions depending on either $x - a_0 t$ or $x + a_0 t$ only, as travelling or stationary waves. Now, if we restrict ourselves to waves moving to the right so that $\psi(x + a_0 t) \equiv 0$, the function $\phi(x - a_0 t)$ satisfies a component of eq. (2.7), namely

$$\rho'_t + a_0 \rho'_x = 0. \tag{2.9}$$

(Of course, it satisfies eq. (2.7) too). Eq. (2.9) is the simplest linear wave equation with solution $\phi(x - a_0 t)$ assuming the initial value $\phi(x)$ and giving at later time the same profile translated as a whole to the right a distance $a_0 t$, without any change in form.

Eq. (2.7) was derived on the assumption that the perturbations in pressure etc. were infinitesimally small. This is not generally true if the agency producing the wave releases large energy or momentum. This is the case, for example, with an explosion, or a relatively fast piston motion. The system (2.3)–(2.5) in essentially two variables (ρ and v, say) is nonlinear and difficult to handle in complete generality. Nevertheless there has been considerable analytic interest in this system. In particular, some progress can be made by seeking simple wave solutions such that one of the dependent variables is a function of the other. This procedure is due originally to Earnshaw (1858). Rewriting eqs. (2.3)–(2.4) in terms of v and ρ

only, by introducing the square of the speed of sound, $a^2 = (\partial p/\partial \rho)_{S=S_0} = k\gamma\rho^{\gamma-1}$, where k is a constant, we have

$$\rho_t + v\rho_x + \rho v_x = 0, \tag{2.10}$$

$$v_t + vv_x + \frac{a^2(\rho)}{\rho}\rho_x = 0. \tag{2.11}$$

Now, we assume that $v = V(\rho)$ so that (2.10)–(2.11) become

$$\rho_t + (V + \rho V')\rho_x = 0, \tag{2.12}$$

$$\rho_t + \left(V + \frac{a^2}{\rho V'}\right)\rho_x = 0. \tag{2.13}$$

Here a prime denotes differentiation with respect to ρ. This system of linear algebraic equations in ρ_t and ρ_x has a non-trivial solution provided the determinant of the coefficient matrix vanishes so that

$$V' = \pm\frac{a}{\rho} = \pm a_0\left(\frac{\rho}{\rho_0}\right)^{(\gamma-1)/2}\frac{1}{\rho}. \tag{2.14}$$

The system (2.10)–(2.11) then reduces to one of the equations

$$\rho_t + (V \pm a)\rho_x = 0, \tag{2.15}$$

where

$$V(\rho) = \int_{\rho_0}^{\rho}\frac{a(\rho)}{\rho}\mathrm{d}\rho = \frac{2}{\gamma-1}\{a(\rho) - a_0\}. \tag{2.16}$$

Restricting attention to waves moving to the right and choosing, therefore, the plus sign in eq. (2.15), we now write the corresponding equation for v. This follows easily from multiplying eq. (2.12) or eq. (2.13) by $V'(\rho)$ and writing the result in terms of v via eq. (2.14). Thus, we obtain

$$v_t + \left(a_0 + \frac{\gamma+1}{2}v\right)v_x = 0. \tag{2.17}$$

The 'simplicity' of the simple wave solution given by eqs. (2.15)–(2.16) or eqs. (2.16)–(2.17) does not arise from any meddling with nonlinearity; these equations are typically nonlinear. The mathematical problem has been reduced to solving an initial value problem for the single first order nonlinear partial differential equation (2.17) consistent with the intermediate integral (2.16) relating v and $a(\rho)$, and hence v and ρ. We note in passing that this argument has been extended to an nth order system of homogeneous PDEs by Schindler (1970). (See also Levine (1972) and Rott (1978).)

If we compare eq. (2.9) with eq. (2.17), we observe that the propagation speed is a_0, a constant, for the former, while it is $a_0 + \frac{1}{2}(\gamma + 1)v$, a function of the dependent variable v itself, for the latter. If an arbitrary profile, however smooth, is chosen initially for v, it is well known and can be graphically checked by drawing the characteristics in the x–t plane that the solution of eq. (2.17), after a certain time, ceases to be single valued. The parts of the initial profile with higher values of v travel faster than those with lower values so that, in due course, in the compressive parts of the initial profile we have three values of the solution, which is impossible. Physically, what transpires is that a shock is formed at a point of the profile where the gradients are large. In a thin neighbourhood of this point, due to the prevalence of large gradients, irreversible thermodynamic processes such as viscosity and heat conduction which were ignored in the derivation of eq. (2.17) intervene. The steepening gradients are eased and a certain balance is struck. The shock with a 'small' thickness then heads the smooth parts of the profile. The details of the shock formation and its subsequent decay in the framework of (the non-viscous and non-heat-conducting equation) (2.17) may be found in Whitham (1974).

Thus, the model equation (2.17) is inadequate to describe flows with shocks and therefore it must be improved upon to include the neglected effects of viscosity and heat conduction. (This was indeed the way it was done in the early stages of the evolution of the topic.) In a heuristic way, eq. (2.17) was 'embedded' with viscosity, so that we obtain

$$v_t + \left(a_0 + \frac{\gamma + 1}{2} v \right) v_x = \frac{\delta}{2} v_{xx} \tag{2.18}$$

(see Cole (1951)). Here δ is a small parameter. This equation can be transformed into eq. (2.1) by a simple change of variables. Eq. (2.1) represents a simple $(1 + 1)$-dimensional model, combining a nonlinear convective term and a small linear viscous term.

2.2 Derivation of Burgers' equation

The Burgers equation and its generalised forms relevant to various physical circumstances have been derived by several investigators using perturbation methods and multiple scales. Here we follow an order-of-magnitude argument due to Lighthill (1956), which leads in the process also to a coupled system of two equations intermediate between the Navier–Stokes equations and the Burgers equation. This system has some intrinsic interest, since its left hand sides are exactly those of one-dimensional isentropic gas dynamic equations (rather than the simple wave form as in the left side of eq. (2.18)), and the right side in one of the equations contains linearised viscous and heat-conduction terms. Thus, starting with the plane Navier–Stokes equations and appropriately grouping various terms (in a certain fashion) we have

$$\frac{\partial\rho}{\partial t}+\rho\frac{\partial v}{\partial x}+\left(v\frac{\partial\rho}{\partial x}\right)=0, \tag{2.19}$$

$$\frac{\partial v}{\partial t}+\left(v\frac{\partial v}{\partial x}\right)+\frac{1}{\rho}\frac{\partial p}{\partial x}=\left[\frac{\frac{4}{3}\mu_0+\mu_{v_0}}{\rho_0}\frac{\partial^2 v}{\partial x^2}\right]$$
$$+\left\{\frac{1}{\rho}\frac{\partial}{\partial x}\left(\left(\frac{4}{3}\mu+\mu_v\right)\frac{\partial v}{\partial x}\right)-\frac{\frac{4}{3}\mu_0+\mu_{v_0}}{\rho_0}\frac{\partial^2 v}{\partial x^2}\right\}, \tag{2.20}$$

$$\frac{1}{\gamma-1}\frac{\mathrm{D}p}{\mathrm{D}t}-\frac{\gamma}{\gamma-1}\frac{p}{\rho}\frac{\mathrm{D}\rho}{\mathrm{D}t}$$
$$=\left\{\left(\frac{4}{3}\mu+\mu_v\right)\left(\frac{\partial v}{\partial x}\right)^2\right\}+\left[k_0\frac{\partial^2 T}{\partial x^2}\right]+\left\{\frac{\partial}{\partial x}\left(\frac{k\partial T}{\partial x}\right)-k_0\frac{\partial^2 T}{\partial x^2}\right\},$$
$$\mathrm{D}=\frac{\partial}{\partial t}+v\frac{\partial}{\partial x}. \tag{2.21}$$

Here μ is the viscosity coefficient equal to the ratio of shear stress to rate of shear, μ_v is the bulk viscosity, and k is the thermal conductivity. T stands for temperature. The suffix 0 denotes values in the undisturbed condition. The constant coefficients k_0, μ_0 and μ_{v_0} are known to be small. The above grouping needs some explanation. The unbracketed terms are the largest and lead to the linearised equation (2.7). If V_0 and a_0 denote a characteristic particle velocity in the wave and the undisturbed value of the speed of sound respectively, then the terms in round brackets are of the order of V_0/a_0 and those in square brackets are of the order $v\omega/a_0^2$ as compared to the unbracketed ones, respectively. The terms in curly brackets are perturbations over the linearised form of viscosity and heat conduction and therefore smaller than those without brackets by an order $(v\omega/a_0^2)\,(V_0/a_0)$. To verify these statements, we may use the linear solution

$$\frac{v}{a_0}=\frac{\rho'}{\rho_0}=\frac{p'}{p_0}=\phi\left(\omega\left(t-\frac{x}{a_0}\right)\right),\quad\text{say,} \tag{2.22}$$

where ω is a typical frequency of the wave, so that

$$\frac{v\rho_x}{\rho_t} \sim \frac{V_0}{a_0}, \tag{2.23}$$

$$\frac{\frac{4}{3}\mu_0 + \mu_{v0}}{\rho_0}\frac{\rho_{xx}}{\rho_t} \sim \frac{\nu\omega}{a_0^2}. \tag{2.24}$$

The coefficients of viscosity and heat conduction are of (small) order $\nu = \mu_0/\rho_0$, the kinematic viscosity. Lighthill has shown that, in the audible sound range, that is, for frequencies between thresholds of hearing and pain, the amplitudes of velocities V_0/a_0 and $\nu\omega/a_0^2$ are of the same order so that terms in both round and square brackets may be retained. The lowest terms in eq. (2.21), namely $p_t \sim a_0 p_x$ and $\gamma\rho_t \sim \gamma a_0 \rho_x$, combine to give entropy changes of (small) order $\nu\omega/a_0^2$ so that the term $k_0 T_{xx}$ is first replaced by density change using the equation of isentropy and the equation of state. The latter then is expressed in terms of velocity change via the linearised form of the left side of (2.21) and the linearised (wave) equation for v. The result is

$$k_0 T_{xx} \sim \frac{k_0 a_0}{c_p} v_{xx}. \tag{2.25}$$

In the process, terms of order $(V_0/a_0)(\nu\omega/a_0^2)$ are neglected. (See Rudenko and Soluyan (1977) for a slightly different derivation.) The linear (time derivative) parts of the convective term in the left side of eq. (2.21) can be approximated by

$$-\frac{a_0}{\gamma - 1}\left[p_x - \frac{\gamma p_0}{\rho_0}\left(\frac{\rho}{\rho_0}\right)^{\gamma-1}\rho_x\right]. \tag{2.26}$$

Here we have used the fact that the assumption of isentropy in the derivation will introduce an error no greater than $(V_0/a_0)(\nu\omega/a_0^2)$ or $(\nu\omega/a_0^2)^2$. Finally, equating (2.25) and (2.26) and eliminating p_x with the help of eq. (2.20), we have, to the order of our approximation,

$$v_t + vv_x + a_0^2\left(\frac{\rho}{\rho_0}\right)^{\gamma-1}\frac{1}{\rho}\rho_x = \left[\frac{\frac{4}{3}\mu_0 + \mu_{v0}}{\rho_0} + \frac{(\gamma - 1)k_0}{\rho_0 c_p}\right]v_{xx}. \tag{2.27}$$

Eqs. (2.19) and (2.27) in v and ρ can be written in a more familiar form in terms of v and the speed of sound,

$$a = a_0\left(\frac{\rho}{\rho_0}\right)^{(\gamma-1)/2}. \tag{2.28}$$

The desired intermediate system is

$$v_t + vv_x + \frac{2}{\gamma - 1} a a_x = \delta v_{xx}, \tag{2.29}$$

$$a_t + va_x + \frac{\gamma - 1}{2} a v_x = 0, \tag{2.30}$$

where

$$\delta = \left(\frac{\frac{4}{3}\mu_0 + \mu_{v0}}{\rho_0} + \frac{\gamma - 1}{\rho_0} \frac{k_0}{c_p} \right) \tag{2.31}$$

may be called diffusivity of sound, 'being that combination of different diffusivities which affects acoustic attenuation'.

The system (2.29)–(2.30) may be referred to as a set constituting generalised Burgers equations for which the left sides equated to zero give the full reversible isentropic system (2.10)–(2.11) when it is expressed in terms of v and a. The right side of eq. (2.29) represents the linearised approximation to the effects of diffusion, to the order of approximation such that the ratios of terms ignored to the largest terms retained are of order $(V_0/a_0)\,(\nu\omega/a_0^2)$ or $(\nu\omega/a_0^2)^2$. This system has not received sufficient attention possibly because of the analytic difficulties it poses (see Itaya (1976)). Its stationary-shock-like solutions depending on $x - Ut$, say, may be easily obtained by solving the ordinary nonlinear differential equation (84) of Lighthill (1956), at least, numerically. The linearised form of this system, assuming $V_0/a_0 \ll \nu\omega/a_0^2$, is

$$\begin{aligned} v_t + \frac{2}{\gamma - 1} a_0 a_x &= \delta v_{xx}, \\ a_t + \frac{\gamma - 1}{2} a_0 v_x &= 0, \end{aligned} \tag{2.32}$$

which can be combined to give

$$v_{tt} - a_0^2 v_{xx} = \delta v_{xxt}. \tag{2.33}$$

Eq. (2.33) has the elementary uni-directional plane wave solution $v = \bar{v}_0 e^{i(\omega t - kx)}$ (with $\bar{v}_0$ a constant) if

$$k^2 = \frac{\omega^2}{a_0^2 + i\delta\omega}$$

or

$$k \sim \frac{\omega}{a_0} - \frac{i}{2}\delta\frac{\omega^2}{a_0^3}$$

(by binomial expansion if once again terms of order $(\nu\omega/a_0^2)^2$ are neglected). Thus, we have

$$v = \bar{v}_0 e^{-\alpha x} e^{i\omega(t - x/a_0)} \tag{2.34}$$

where $\alpha = \delta\omega^2/2a_0^3$. The coefficient α describes the rate at which the amplitude of the monochromatic wave with frequency ω attenuates due to diffusive damping. If we consider waves generated at the boundary $x = 0$ according to the condition

$$v_0(t) = \int_{-\infty}^{\infty} e^{i\omega t} F(\omega) d\omega, \tag{2.35}$$

in terms of a Fourier transform of a function $F(\omega)$, then the solution can be expressed by superposition as

$$\begin{aligned} v &= \int_{-\infty}^{\infty} e^{i\omega(t - x/a_0)} e^{-(1/2)\delta\omega^2(x/a_0^3)} F(\omega)\, d\omega \\ &= \frac{1}{2\pi}\int_{-\infty}^{\infty} e^{i\omega(t - x/a_0)} e^{-(1/2)\delta\omega^2(x/a_0^3)} \left(\int_{-\infty}^{\infty} e^{-i\omega s} v_0(s)\, ds \right) d\omega \\ &= \frac{1}{(2\pi\delta x/a_0)^{1/2}} \int_{-\infty}^{\infty} v_0(s) \exp\left[-\frac{(a_0 t - x - a_0 s)^2}{2\delta(x/a_0)} \right] a_0\, ds \end{aligned} \tag{2.36}$$

by rearranging terms and using Fourier inversion. This form of the solution brings out clearly the diffusive character of the wave which is now centred on the line $s = t - x/a_0$ and has diffusivity $(\delta x/a_0)^{1/2} \sim (\delta t)^{1/2}$ – hence the appropriateness of calling the coefficient δ the diffusivity of sound.

We now return to the system (2.29)–(2.30) and attempt further simplification. If we introduce the Riemann invariants r and s, this system becomes

$$r_t + (a + v) r_x = \tfrac{1}{2}\delta(r_{xx} - s_{xx}), \tag{2.37}$$

$$s_t - (a - v) s_x = \tfrac{1}{2}\delta(s_{xx} - r_{xx}), \tag{2.38}$$

where

$$r = \frac{a}{\gamma - 1} + \tfrac{1}{2}v, \quad s = \frac{a}{\gamma - 1} - \tfrac{1}{2}v, \tag{2.39}$$

so that

$$a + v = \tfrac{1}{2}(\gamma + 1) r + \tfrac{1}{2}(\gamma - 3) s,$$

$$a - v = \tfrac{1}{2}(\gamma - 3) r + \tfrac{1}{2}(\gamma + 1) s.$$

Further approximation to arrive at the Burgers equation consists in assuming that the Riemann invariant s (even in this dissipative case) is a constant equal to its value s_0 in the undisturbed state. With this approximation eq. (2.37) becomes an equation for the single unknown r, namely

$$r_t + [\tfrac{1}{2}(\gamma+1)r + \tfrac{1}{2}(\gamma-3)s_0]r_x = \frac{\delta}{2} r_{xx}. \tag{2.40}$$

To lowest order (that is, if terms of order V_0/a_0 and $\nu\omega/a_0^2$ are neglected) both a and v satisfy the linear wave eq. (2.9) as do r and s. We have, therefore,

$$\frac{\partial}{\partial t}\left(\frac{\partial r}{\partial x}\right) \approx -a_0 \frac{\partial}{\partial x}\left(\frac{\partial r}{\partial x}\right). \tag{2.41}$$

If we now linearise eq. (2.38) by neglecting terms of order $(V_0/a_0)^2$ and use eq. (2.41), we can easily verify that the quantity

$$\frac{s - s_0}{a_0} = \frac{\delta}{4a_0^2} \frac{\partial r}{\partial x} \tag{2.42}$$

satisfies the resulting equation, with an error of still higher order. No terms of order V_0/a_0 have been neglected in dealing with r so that the (non-dimensional) excess of r over its undisturbed value $(r - r_0)/a_0$ is of order V_0/a_0, while eq. (2.42) shows that $(s - s_0)/a_0$ is of order $(\nu\omega/a_0^2)(V_0/a_0)$ so that $\frac{1}{2}\delta s_{xx}$ in eq. (2.37) can be neglected. We emphasise that the quadratic terms in V_0/a_0 and $\nu\omega/a_0^2$ have been neglected in deriving eq. (2.42).

Broer and Schuurmans (1970) posed the question of the accuracy of the approximation $s = s_0$ in eqs. (2.37) and (2.38), with compatible initial conditions $r(x, 0) = F(x)$, $s(x, 0) = s_0$. In particular the ability of the (approximating) Burgers equation to give reasonably accurate solutions for a finite time depending on the initial function $f(x)$ was discussed. However, no analysis of the system (2.37)–(2.38) was carried out. Instead, another linear system close enough to this system was solved. The question probably can only be answered by solving such a problem numerically and comparing it with the exact solution of the Burgers equation subject to (corresponding) compatible initial conditions. Now, to bring eq. (2.40) to the familiar form (2.1), we introduce the variables

$$\begin{aligned} u &= \tfrac{1}{2}(\gamma+1)(r - r_0) \\ &= \tfrac{1}{2}(\gamma+1)r + \tfrac{1}{2}(\gamma-3)s_0 - a_0 \approx a + v - a_0, \\ X &= x - a_0 t, \end{aligned} \tag{2.43}$$

where $r_0 = s_0 = a_0/(\gamma - 1)$, and the approximation $s = s_0$ has been used in the definition of $(a+v)$. Eq. (2.40) now becomes

$$u_t + uu_X = \frac{\delta}{2} u_{XX}. \tag{2.44}$$

The quantity u is essentially the excess Riemann invariant or more physically (approximate) excess wavelet speed which causes convective steepening of the wave. The co-ordinate $X = x - a_0 t$ is a spatial coordinate moving with the undisturbed sound speed in the direction of the wave. Henceforth, we let x denote the moving co-ordinate X in the Burgers eq. (2.44).

It is of some interest to compare the above derivation with that of Karpman (1975) and Rudenko and Soluyan (1977). The first part of their reduction is similar to the derivation of the intermediate system (2.29)–(2.30). In the second part, they express p and ρ as

$$p(x,t) = p(v) + \psi(x,t),$$
$$\rho(x,t) = \rho(v) + \phi(x,t),$$

where the functions $p(v)$ and $\rho(v)$ are precisely those arising from the simple wave assumption for the non-dissipative system (cf. Eq. (2.16)) and $\psi(x,t)$ and $\phi(x,t)$ are perturbations of second order such that the corresponding solutions remain as 'close' to simple waves as possible. It turns out that ψ is proportional to the spatial derivative of v, and ϕ is equal to zero, to the required approximation leading to the Burgers equation. The resulting solution is referred to as a quasi-simple wave. This term was first introudced by Courant and Friedrichs (1948) to describe spherical and cylindrical shock wave solutions depending on r/t only, and take into account the lower order (undifferentiated) terms $2u/r$ or u/r in the equation of continuity (see also Seshadri and Sachdev (1977) for quasi-simple waves for multi-dimensional systems). Such quasi-simple waves, though enjoying some properties of the simple waves, differ quite significantly. In the present context, solutions are sought which are functions of one of the dependent variables as well as its derivative to account for terms involving derivatives higher than first. Moreover, the treatment here is approximate.

2.3 Historical background and transformations

Eq. (2.44), in its present one-dimensional form, was first mooted by Bateman (1915), who found its steady solutions descriptive of certain viscous flows. It was later proposed by Burgers (1940) as one of a class of equations to describe mathematical models of turbulence. In the context of gas dynamics, it was discussed by Hopf (1950) and Cole (1951).

The pre-eminence of this equation rests on its exact linearisability, through the nonlinear transformation

$$u = -\,\delta(\log\phi)_x, \tag{2.45}$$

to the standard heat equation. Moreover, the initial condition for eq. (2.44) transforms in a direct simple manner into the initial condition for the heat equation. It appears that the relationship of the Burgers equation with the heat equation was first noted in the book by Forsyth (1906), without any allusion to a physical problem. Hopf remarks in a footnote to his paper that 'the reduction of (1) [Burgers' equation] to the heat equation was known to me since the end of 1946. However, it was not until 1949 that I became sufficiently acquainted with the recent developments of fluid dynamics to be convinced that a theory of (1) [Burgers' equation] could serve as an instructive introduction into some of the mathematical problems involved.' The transformation in a fluid dynamic context appeared first in a technical report by Lagerstrom, Cole and Trilling (1949). It was later published by Cole (1951) in his treatment of some 'aerodynamic problems' through Burgers equation.

This transformation which is of Bäcklund type (Rogers and Shadwick, 1982) has had far-reaching influence in inducing researchers to extend it to generalised Burgers equations or to nonlinear dispersive equations of Korteweg–de Vries type – we shall have occasion to refer to these matters subsequently. Apparently, the inspiration for this transformation came from a similar transformation for the nonlinear ordinary differential equation of first order

$$u' = f(x) + g(x)u + h(x)u^2, \tag{2.46}$$

known as the generalised Riccati equation. Thus,

$$u = -\frac{1}{h(x)}\frac{\mathrm{d}}{\mathrm{d}x}(\ln\phi) \tag{2.47}$$

changes eq. (2.46), with a further simple transformation, into the canonical linear second order equation

$$\Phi'' + p(x)\Phi = 0, \tag{2.48}$$

where $p(x)$ is a specified combination of the functions f, g and h. Indeed, Burgers (1950) arrived at this transformation by seeking similarity solutions $u = t^{-1/2}S(z)$, $z = (2\delta t)^{-1/2}x$ for his equation. The similarity form of the equation comes out to be of Riccati type. This equation also appears in the similarity forms of more general equations of Burgers type.

The fortunate nature of Burgers' equation is that, after the transformation (2.45), the equation can be integrated with respect to x once, leading again to a second order equation, the heat equation. It is convenient to apply eq. (2.45) to eq. (2.44) in two steps. We first put

$$u = \psi_x \tag{2.49}$$

in eq. (2.44) and integrate with respect to x, ignoring the 'function' of integration. We have

$$\psi_t + \tfrac{1}{2}\psi_x^2 = \frac{\delta}{2}\psi_{xx}. \tag{2.50}$$

Next the transformation

$$\psi = -\delta(\ln \phi) \tag{2.51}$$

changes eq. (2.50) into

$$\phi_t = \frac{\delta}{2}\phi_{xx}. \tag{2.52}$$

It is interesting to compare this reduction to an analogous one for the K–dV equation (2.2). Whitham (1974) introduced the 'natural' generalisation of eqs. (2.49) and (2.51),

$$\sigma u = 12(\ln \phi)_{xx}, \tag{2.53}$$

(numerical factors in eqs. (2.51) and (2.53) are important and specific), and arrived at the equation

$$\phi(\phi_t + \phi_{xxx})_x - \phi_x(\phi_t + \phi_{xxx}) + 3(\phi_{xx}^2 - \phi_x\phi_{xxx}) = 0, \tag{2.54}$$

which is a further 'non-linearisation' of the original equation! But, in the process, the transformation introduces a certain 'order' so that all terms are uniformly of second degree. Hirota (1971) has extended this kind of transformations to other nonlinear dispersive equations. The initial value problem for the K–dV equation has been 'linearised' via the linear integral equation of Gelfand–Levitan type by the method of inverse scattering (see Whitham (1974)). Whether, conversely, the inverse scattering techniques can be extended to generalised Burgers equations remains to be investigated. Attempts to extend Hopf–Cole like transformations to these equations will be discussed in chapter 3.

Burgers' equation (as well as the Hopf–Cole transformation and the heat equation) enjoys several properties which make it a popular model for the discussion of modern techniques of the transformation theory of nonlinear partial differential equations. The transformations

(1) the shift of origin

$$x - x_0 \to x, \quad t - t_0 \to t; \quad u \to u, \quad \phi \to \phi, \tag{2.55}$$

where x_0 and t_0 are arbitrary, independent constants,

(2) the change of scale

$$\frac{x}{\alpha} \to x, \quad \frac{t}{\alpha^2} \to t; \quad \alpha u \to u, \quad \beta\phi \to \phi, \tag{2.56}$$

where α and β are arbitrary independent scale factors, and

(3) the Galilean transformation

$$x - Ut \to x, \quad t \to t; \quad u - U \to u \tag{2.57}$$

with real U leave eq. (2.44) invariant. Eqs. (2.45) and (2.52) are invariant under (2.55) and (2.56). These transformations help in generating new solutions; however, since, for real values of the constants in (2.55)–(2.57), the solutions obtained by these transformations will have the same form as the original ones, these solutions may be referred to as equivalent or isomorphic, and hence physically not distinct.

In recent years, the Lie–Bäcklund transformation theory of nonlinear partial differential equations has been developed extensively to reduce the number of independent variables by one. For equations with two independent variables, it amounts to reducing the given system of partial differential equations to ordinary differential equations through the so-called similarity variables, introducing considerable simplification in their solution. This theory, first developed by Lie (1891), and later by Ovsiannikov (1962, 1982), was made popular by the work of Bluman and Cole (1969), who applied it to the heat equation to identify the similarity variables and hence find all the similarity solutions. Until recently most (physically) important similarity solutions were found by intuitive and dimensional arguments. One might, for example, refer to the well-known blast wave solution (Sedov, 1946, Taylor, 1950), and the converging shock wave solution (Guderley, 1942). However, there are some special situations for which only the theory of finite and infinitesimal transformations can identify the similarity variables. This happens when the so-called characteristic equations in the latter method determining the similarity variables assume special values for some of the parameters occurring in them, and their integrals lead to similarity variables involving the logarithm of one of the independent variables. Such similarity solutions would be hard to discover by intuitive arguments alone (see, for example, Sachdev and Reddy 1982).

In fact, this topic has been discussed quite thoroughly by Ames (1972), who covers in detail aspects relevant to Burgers' equation. We, therefore, content ourselves with a brief mention of some of the results. The invariance of the solution surface $u = \theta(x, t)$ and the Burgers equation (2.44) under the infinitesimal transformations

$$\begin{aligned} x' &= x + \varepsilon X(x, t, u) + O(\varepsilon^2), \\ t' &= t + \varepsilon T(x, t, u) + O(\varepsilon^2), \\ u' &= u + \varepsilon U(x, t, u) + O(\varepsilon^2), \end{aligned} \tag{2.58}$$

leads to the characteristic equations of a certain Lagrange equation,

$$\frac{\mathrm{d}x}{X} = \frac{\mathrm{d}t}{T} = \frac{\mathrm{d}\theta}{\theta}, \tag{2.59}$$

whose integrals determine the similarity form of the solution. The functions $X, T,$ and U are governed by a coupled system of determining partial differential equations in these variables, arising out of the invariance of the Burgers equation. On most occasions, it is not possible to find a general solution of these equations. However, each special solution, when substituted in eq. (2.59), leads to a similarity form of the original equation (which, here, is the Burgers equation). The similarity form of the solution can be obtained without reference to the boundary conditions, though the method as a whole requires the invariance of the boundary conditions under eqs. (2.58) as well. The whole procedure, particularly if one has to deal with a simultaneous system of nonlinear partial differential equations, is quite cumbersome.

The special cases deduced by Ames (1972) for the Burgers equation (2.44) with $\delta = 2$ lead to the following similarity solutions (the factor $\delta/2$ can, in fact, be scaled out of the equation).

(a) $$u = \frac{1}{(2t + m)^{1/2}} f(\eta), \quad \eta = \frac{x}{(2t + m)^{1/2}} \tag{2.60}$$

where f is governed by the nonlinear equation

$$f'' + f'(\eta - f) + f = 0. \tag{2.61}$$

We shall see later on that this corresponds to the so-called single hump solution (see sec. 2.6). Its solution can be explicitly written.

(b) $$u = \frac{1}{\eta} + (t+d)^{-1} f(\eta), \quad \eta = \frac{t+d}{x+R}, \tag{2.62}$$

where f satisfies the nonlinear equation

$$\eta^2 f'' + 2\eta f' + ff' = 0. \tag{2.63}$$

The solution of the latter equation is

$$f = a_2 \tanh\left[\frac{a_2}{2}(a_3 - \eta^{-1})\right]. \tag{2.64}$$

In (a) and (b) above, m, d, R, a_2 and a_3 are arbitrary constants. Ames (1972) also determined the one-parameter group of finite transformations under which the Burgers equation remains invariant, and arrived at the similarity solution

(c) $$u = -1 + \frac{1}{t^{1/2}} f(\eta), \quad \eta = \frac{x+t}{t^{1/2}}, \tag{2.65}$$

where f satisfies the equation

$$f'' - ff' + \tfrac{1}{2}\eta f' + \tfrac{1}{2} f = 0.$$

Its first integral

$$f' - \tfrac{1}{2} f^2 + \tfrac{1}{2}\eta f = \text{constant} \tag{2.66}$$

is a special form of the Riccati equation.

Finally, we also refer to the work of Chester (1977) who has discussed continuous transformations from a somewhat different viewpoint. He has also found the similarity form of the solution for Burgers' equation, which is a slight generalisation of (2.60).

We insert some remarks concerning the physical similitude of solutions of eq. (2.44). If l denotes the typical length of the wave, u_0 its typical amplitude and δ the coefficient of viscous diffusion, then (by dimensional considerations) we have the non-dimensional variables

$$U = \frac{ul}{\delta}, \quad \tau = \frac{\delta t}{l^2}, \quad \xi = \frac{x}{l}, \quad R_0 = \frac{u_0 l}{\delta} \tag{2.67}$$

where R_0 is a non-dimensional parameter, called the Reynolds number, so that the solution

$$U = \frac{ul}{\delta} = F(R_0, \tau, \xi) \tag{2.68}$$

depends on R_0 as well as on τ and ξ. R_0 is a measure of the (initial) nonlinearity as opposed to viscosity. To compare two solutions, say, with the same l but different viscosities, we would have the same value of R_0, U and τ if $t_1\delta_1 = t_2\delta_2$ and $u_1/\delta_1 = u_2/\delta_2$. In the linearised form of the Burgers equation, the nonlinearity parameter R_0 drops out.

2.4 The pure initial value problem

The connection (2.45) of the solution u of Burgers' equation (2.44) with that of the heat equation makes the pure initial value problem for the former a relatively simple matter. Suppose we are given

$$u(x,0) = f(x), \quad -\infty < x < \infty; \tag{2.69}$$

then, writing the transformation (2.45) in an integrated form (ignoring the function of integration), we have

$$\phi(x,t) = \exp\left[-\frac{1}{\delta}\int^{x} u(\xi,t)\mathrm{d}\xi\right], \tag{2.70}$$

so that the appropriate initial condition for the heat equation (2.52) is

$$\begin{aligned}\phi(x,0) &= \exp\left[-\frac{1}{\delta}\int_0^{x} u(\xi,0)\mathrm{d}\xi\right] \\ &= \exp\left[-\frac{1}{\delta}\int_0^{x} f(\xi)\mathrm{d}\xi\right] = \Phi(x),\end{aligned} \tag{2.71}$$

say, choosing the lower limit zero for x in the definition of $\Phi(x)$.

Thus we have the following precise statement of the existence and uniqueness theorem for eqs. (2.44) and (2.69) due to Hopf (1950). Suppose $f(x)$ is integrable in every finite x interval and

$$\int_0^{x} f(\xi)\mathrm{d}\xi = o(x^2) \tag{2.72}$$

for $|x|$ large. Then

$$u(x,t) = \frac{\displaystyle\int_{-\infty}^{\infty} \frac{x-\xi}{t}\exp\left[-\frac{1}{\delta}\mathrm{F}(x,\xi,t)\right]\mathrm{d}\xi}{\displaystyle\int_{-\infty}^{\infty} \exp\left[-\frac{1}{\delta}\mathrm{F}(x,\xi;t)\right]\mathrm{d}\xi}, \tag{2.73}$$

where

$$F(x,\xi,t)=\frac{(x-\xi)^2}{2t}+\int_0^{\xi} f(\eta)\mathrm{d}\eta, \tag{2.74}$$

is a regular solution of eq. (2.44) in the half plane $t>0$ that satisfies the initial condition

$$\int_0^x u(\xi,t)\mathrm{d}\xi \to \int_0^a f(\xi)\mathrm{d}\xi \quad \text{as } x\to a, \quad t\to 0 \tag{2.75}$$

for every a. If, in addition, $f(\xi)$ is continuous at $x=a$, then

$$u(x,t)\to f(a) \quad \text{as } x\to a, \quad t\to 0. \tag{2.76}$$

A solution of eq. (2.44) which is regular in the strip $0<t<T$ and which satisfies (2.75) for each value of the number a necessarily coincides with (2.73) in the strip.

In fact, the solution of the heat equation (2.52) subject to (2.71) exists and is unique only if the asymptotic condition (2.72) on its initial behaviour is satisfied; indeed, there is the counter-example of Tychnov (see Copson (1975)) to show that the uniqueness is violated if (2.72) does not hold. The explicit solution of (2.52) and (2.71), through the principle of linear superposition via the source or fundamental solution, is given by

$$\begin{aligned}\phi &= \frac{1}{(2\pi\delta t)^{1/2}}\int_{-\infty}^{\infty}\Phi(\xi)\exp\left[-\frac{(x-\xi)^2}{2\delta t}\right]\mathrm{d}\xi \\ &= \frac{1}{(2\pi\delta t)^{1/2}}\int_{-\infty}^{\infty}\exp\left\{-\frac{1}{\delta}\left[\frac{(x-\xi)^2}{2t}+\int_0^{\xi} f(\eta)\mathrm{d}\eta\right]\right\}\mathrm{d}\xi.\end{aligned} \tag{2.77}$$

The solution (2.77) is the same as the denominator of (2.73) except for the factor $(2\pi\delta t)^{-1/2}$. This factor disappears when the transformation (2.45) is used to get $u(x,t)$. Hopf has established the existence theorem by proving the corresponding theorem for the heat equation. It is necessary to have the denominator of (2.73) positive to avoid singularities in u. The solution (2.73) of the Burgers equation satisfies the initial condition (2.75) at every point $x=a$ in the integral sense; this is satisfied pointwise if the initial function $f(x)$ is continuous everywhere in the interval.

If the condition (2.72) is weakened so that we have

$$\int_0^x f(\xi)\mathrm{d}\xi = O(x^2) \tag{2.78}$$

for large $|x|$, then the solution of the initial value problem exists and is regular only in a finite time interval. This is illustrated by the solution $u = x/(t - T)$, for which eq. (2.78) holds. This solution blows up at a finite t equal to T.

It is clear that the integrals in the solution (2.73) of the initial value problem cannot, in general, be expressed in a closed form for an arbitrary initial function $f(x)$. Lighthill (1956) used the method of steepest descent and graphical construction of characteristics etc. to find the approximate solutions of several physical problems involving shocks. Many important results regarding the decay of shocks, the displacement of shocks due to diffusion, and interiors of shocks and their confluence etc. were deduced. Some of this work has been succinctly described by Whitham (1974), and followed later by Rudenko and Soluyan (1977). Since one of the purposes of this chapter is to understand clearly the analytic structure of the physically important problems to aid the construction of approximate analytic or numerical solutions of more general Burgers equations, we discuss those solutions which can be explicitly obtained. Benton and Platzman (1972) have compiled an exhaustive list of solutions of the Burgers equation and have illustrated the physically interesting ones by means of isochronal graphs. Their tables incorporate explicit solutions of the heat equation and the corresponding solutions of the Burgers equation, together with some explanation. We shall show later how most of these solutions may be interpreted as arising out of boundary conditions imposed at $x = 0$. First we consider some important special solutions and their physical significance.

2.5 Stationary solutions and shock structure

The Burgers equation is quasi-linear: it is linear in u_{xx}. Besides, the coefficients of the derivative terms depend only on the dependent variable. Therefore, it admits travelling wave solutions depending on $x - Ut$ only, where U is the (constant) speed of the wave. Obviously, such solutions, if they exist, have the same form for all time. Substituting $\xi = x - Ut$ in eq. (2.44) and looking for shock-like solutions with $u \to u_1, u_2$ as $\xi \to \pm\infty$, where $u_2 > u_1$, we have

$$-Uu_\xi + uu_\xi = \frac{\delta}{2}u_{\xi\xi}. \tag{2.79}$$

Since the solutions of this equation assume constant values u_1, u_2 at $\xi = \pm\infty$, the derivative of u vanishes as $\xi \to \pm\infty$, so that integration of eq. (2.79) gives

$$\tfrac{1}{2}u^2 - Uu + C = \frac{\delta}{2}u_\xi \tag{2.80}$$

where C is a constant. Further, imposition of end conditions at $\xi = \pm\infty$ leads to

$$\tfrac{1}{2}u_1^2 - Uu_1 = \tfrac{1}{2}u_2^2 - Uu_2 = -C. \tag{2.81}$$

Eqs. (2.81) give the values of the constants as

$$U = \tfrac{1}{2}(u_1 + u_2), \quad C = \tfrac{1}{2}u_1 u_2. \tag{2.82}$$

Substituting (2.82) in eq. (2.80), we have

$$(u - u_1)(u_2 - u) = -\delta u_\xi \tag{2.83}$$

which integrates to give

$$\frac{\xi}{\delta} = \frac{1}{u_2 - u_1}\ln\frac{u_2 - u}{u - u_1},$$

or, more explicitly,

$$u = u_1 + \frac{u_2 - u_1}{1 + \exp\dfrac{u_2 - u_1}{\delta}(x - Ut)}, \quad U = \frac{u_1 + u_2}{2}. \tag{2.84}$$

The solution (2.84) describes the structure of a uniformly propagating shock with end conditions u_1 and u_2. The velocity of the shock U is the mean of the end velocities and is independent of the shock structure. This, in the terminology of Barenblatt and Zel'dovich (1972), is called a self-similar solution of the first kind, for which the velocity of propagation of the wave is obtained explicitly in terms of the known end conditions. It is also an intermediate asymptotic (see chapter 4) to which a class of solutions arising out of different initial conditions with asymptotically the same end conditions converge as $t \to \infty$. To see this we consider the following initial step conditions which may be thought of as arising from a fast piston motion such that 'the wave form gets away from the piston in a time negligible compared with the time scale of the process of shockwave formation in which we are interested' (Lighthill, 1956):

$$u(x, 0) = f(x) = \begin{Bmatrix} u_1, & x > 0, \\ u_2, & x < 0, \end{Bmatrix} \tag{2.85}$$

where $u_2 > u_1$. The initial conditions in terms of ϕ become

$$\phi(x,0) = \begin{cases} e^{-u_1 x/\delta}, & x>0, \\ e^{-u_2 x/\delta}, & x<0. \end{cases} \tag{2.86}$$

Substituting (2.85) in eq. (2.73) and rearranging suitably, we have

$$u(x,t) = u_1 + \frac{u_2 - u_1}{1 + \left\{ \exp\left[\dfrac{u_2-u_1}{\delta}\left(x - \dfrac{u_1+u_2}{2}t\right)\right]\right\} \dfrac{\int_{-(x-u_1 t)}^{\infty} e^{-y^2/2\delta t}\,dy}{\int_{x-u_2 t}^{\infty} e^{-y^2/2\delta t}\,dy}} \tag{2.87}$$

In the limit as $t\to\infty$ such that $u_1 < x/t < u_2$, the lower limits of the integrals in the denominator of eq. (2.87) are both large and negative so that the ratio of the integrals tends to 1. The solution (2.87) tends to the travelling wave solution (2.84). The latter is also referred to as Taylor shock structure (Taylor, 1910). Here, we have given only one special initial condition such that the solution of the Burgers equation subject to this condition tends in the limit $t\to\infty$ to the travelling wave solution. There can be an infinity of initial conditions with the same end states but 'reasonable' behaviour in the middle which evolve, in the limit $t\to\infty$, to the steady shock structure or the travelling wave solution (2.84), (Barenblatt & Zel'dovich, 1972).

2.6 Single hump solution

The travelling wave solution discussed earlier is a similarity solution of the first kind in the terminology of Barenblatt and Zel'dovich (1972); it can be transformed to appear like a familiar similarity solution. There is also a similarity solution of the second kind whose form can be written by dimensional argument. It appears like an (unsymmetric) single hump with zero values at $x=\pm\infty$. If we produce a compression pulse by moving a piston very rapidly into gas, say, for a distance h and then stopping it, we can define a non-dimensional number, the Reynolds number

$$R = \frac{1}{\delta}\int_{-\infty}^{\infty} u\,dx = \frac{A}{\delta}(\text{say}), \tag{2.88}$$

which has a nice interpretation as the area under the profile of the pulse (a good representation of the product of velocity amplitude and length scale for such a pulse) divided by the coefficient of viscous diffusion. It is a good measure of the ratio of nonlinear convective to (linear) diffusive effects. If we integrate eq. (2.44) with respect to x from $-\infty$ to $+\infty$ and impose the conditions that $u=0$ at both ends, then it immediately follows that R is constant. The existence of such 'stationary' solutions with a fixed 'momentum' R was mentioned by Hopf (1950) who also described the fact that all solutions of the Burgers equation with vanishing conditions at $\pm\infty$ and with a given R will tend asymptotically to the 'stationary' or similarity solution. Eq. (2.88) represents an integral of the Burgers equation.

There are only two dimensional parameters appearing in the problem, namely A and δ, each having the dimensions L^2/T; there are no other dimensional parameters to render length and time separately non-dimensional. Hence the solution takes the similarity form

$$u=\left[\frac{\delta}{t}\right]^{1/2} f\left[\frac{x}{(\delta t)^{1/2}};\frac{A}{\delta}\right]. \tag{2.89}$$

This corresponds to the similarity solution of the heat equation

$$\phi=c_1+c_2\int_{x/(\delta t)^{1/2}}^{\infty} \mathrm{e}^{-y^2/2}\,\mathrm{d}y=c_1+\bar{c}_2\,\mathrm{erfc}\frac{x}{(2\delta t)^{1/2}}, \tag{2.90}$$

which depends on the similarity variable $x/(\delta t)^{1/2}$ only. The constants c_1 and c_2 are chosen such that $u\to 0$ as $x\to\pm\infty$, according to the transformation (2.70) slightly modified as

$$\phi(x,t)=\exp\left(\frac{1}{\delta}\int_x^{\infty} u\,\mathrm{d}x\right). \tag{2.91}$$

The constants c_1 and c_2, therefore, are given by

$$c_1=1,\quad c_2=\frac{\mathrm{e}^R-1}{\sqrt{(2\pi)}}. \tag{2.92}$$

The Hopf–Cole transformation (2.45) then gives

$$\begin{aligned} u&=\left(\frac{\delta}{t}\right)^{1/2}\frac{\mathrm{e}^{-x^2/2\delta t}}{\dfrac{\sqrt{(2\pi)}}{\mathrm{e}^R-1}+\displaystyle\int_{x/(\delta t)^{1/2}}^{\infty}\mathrm{e}^{-y^2/2}\mathrm{d}y} \\ &=\left(\frac{\delta}{t}\right)^{1/2}\frac{\mathrm{e}^{-x^2/2\delta t}}{\dfrac{\sqrt{(2\pi)}}{\mathrm{e}^R-1}+\sqrt{(\pi/2)}\,\mathrm{erfc}\dfrac{x}{(2\delta t)^{1/2}}}. \end{aligned} \tag{2.93}$$

This solution has different forms depending on the magnitude of the Reynolds number R of the initial profile. R, for a given profile, remains constant for all time (see fig. 2.1). When $R \approx 0$, the first term in the denominator of eq. (2.93) dominates the integral term and we have merely the source solution of the diffusion equation, the nonlinear term in the Burgers equation taking no part in the solution. When R is very large, the first term in the denominator is small, but its relative size in comparison to the second term depends on the sign of x. For $x < 0, x/(2\delta t)^{1/2} \equiv \xi \ll -1$, erfc $\xi \approx 2$, and the solution therefore is essentially Gaussian, implying that mainly diffusive effects prevail. Besides, for $x \gg 1, \xi \gg 1$, erfc $\xi \approx 0$, and the region is again diffusive. For $\xi \gtrsim \sqrt{2}$, a good approximation to erfc ξ is $\pi^{-1/2}(\xi)^{-1}\,\mathrm{e}^{-\xi^2}$ so that, for large R,

$$u \sim \frac{x}{t}. \tag{2.94}$$

This is also an exact solution of the Burgers equation in which the diffusive term is trivially computed. This 'inviscid' region is followed by a thin 'shock' layer on the right when the terms e^{-R} and erfc$[x/(2\delta t)^{1/2}]$ in the denominator of eq. (2.93) are comparable. The shock layer has a diffusive forerunner where $\xi \gg \sqrt{2}$ as noted above, and the solution is close to zero there.

Whitham (1974) has discussed the structure of this solution in great detail for large R. There are two transition layers – one the shock layer which in the limit of $R \to \infty$ goes into a discontinuity in u, and the other a transition near $x = 0$, which becomes a discontinuity in the derivative of u at $x = 0$ in

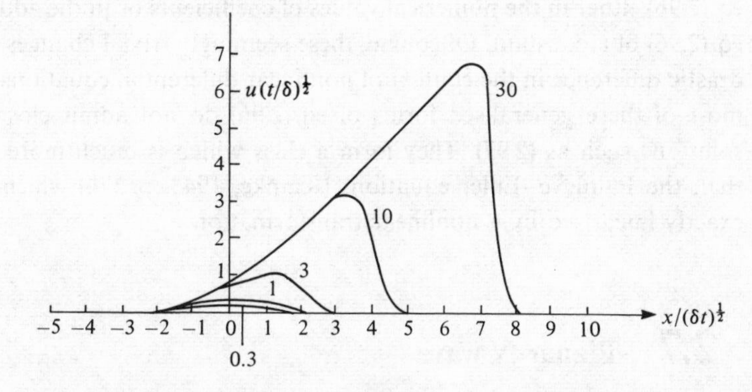

Fig. 2.1. Asymptotic form of the single hump solutions of Burgers equation, with Reynolds numbers 30, 10, 3, 1, 0.3 (from Lighthill (1956)).

the same limit. The shock is located at $x = (2At)^{1/2}$ with the structure given by eq. (2.84).

The similarity solutions, *per se*, arise out of very special, usually singular, initial conditions, but assume importance by virtue of being intermediate asymptotics. For example, Whitham has identified the initial conditions for the single hump solution as a delta function superposed on the undisturbed (constant) value of u. Indeed, there can be many representations of the initial conditions which go into the delta function in some limit, and have a solution evolving into the single hump in the same limit (see, for example, Benton and Platzman (1972), eq. (4.4)).

We conclude the discussion of the single hump solution by noting that the transformation

$$u = \left(\frac{\delta}{t}\right)^{1/2} [H(\xi)]^{-1} \tag{2.95}$$

changes the Burgers equation into the nonlinear ordinary differential equation in the similarity variable $\xi = x/(2\delta t)^{1/2}$,

$$HH'' - 2H'^2 + 2\xi HH' - 2\sqrt{2}H' - 2H^2 = 0, \tag{2.96}$$

where the prime denotes differentiation with respect to ξ. This equation has the exact solution

$$H = \frac{\sqrt{(2\pi)}}{e^R - 1} e^{\xi^2} + \sqrt{(\pi/2)} \cdot e^{\xi^2} \operatorname{erfc} \xi. \tag{2.97}$$

We shall see in chapter 5 that a class of generalised Burgers equations admit similarity solutions governed by equations which differ from eq. (2.96) either in the numerical values of coefficients or in the addition to eq. (2.96) of a constant. Of course, these seemingly trivial changes make a drastic difference in the context of nonlinear differential equations so that most of these generalised forms of eq. (2.96) do not admit closed form solutions such as (2.97). They form a class which is much more general than the Painlevé–Euler equations (Kamke, 1943, p. 574) which can be exactly linearised by a nonlinear transformation.

2.7 Planar N wave

A solution of the Burgers equation which is physically more important than the single hump solution is the so-called N wave. In fact, the N form of the waves which actually arise in nature, say from a spherical explosion or an aircraft, with idealised spherical or cylindrical symmetry, far from the source, carries important geometrical expansion effects due to the shape of the source. These N waves will be referred to as spherical and cylindrical, respectively, and are governed by a generalised Burgers equation to be discussed in the next chapter. The corresponding solution for the plane N waves throws much light on the structure of the solution for the real N waves, which, unfortunately, do not seem to be exactly expressible in a simple analytic form.

The N waves with discontinuous shocks at $|x| = 1$ on the front and the tail, which would emerge from a planar source far away, have the balanced form

$$u = (x, t_i) = \begin{Bmatrix} x & \text{for} |x| < 1, \\ 0 & \text{for} |x| > 1. \end{Bmatrix} \tag{2.98}$$

Evolution of the solution from this initial stage $t = t_i$ to the one when the Taylor shock is well-formed may be obtained analytically (see Crighton and Scott (1979)). Of course, it may also be found numerically. However, after this embryonic stage, there is a solution of the Burgers equation which describes the wave over an infinitely long time during which it decays. We look for a solution of the heat equation, which is even about the node of the wave, $x = 0$, so that the corresponding solution of the Burgers equation, because of the Hopf–Cole transformation (2.45), is odd about $x = 0$, and vanishes at $x = \pm\infty$. Such a solution of the heat equation is

$$\phi = 1 + \left(\frac{t_0}{t}\right)^{1/2} \mathrm{e}^{-x^2/2\delta t}, \tag{2.99}$$

where t_0 is a constant. The corresponding solution of the Burgers equation is

$$u = -\delta \frac{\phi_x}{\phi} = \frac{x/t}{1 + \left(\dfrac{t}{t_0}\right)^{1/2} \mathrm{e}^{x^2/2\delta t}}. \tag{2.100}$$

We define the Reynolds number in the present case as the area under one of the two (equal) lobes divided by δ,

$$R = \frac{A}{\delta} = \frac{1}{\delta} \int_0^\infty u \, \mathrm{d}x = \ln \phi(0, t) = \ln\left[1 + \left(\frac{t_0}{t}\right)^{1/2}\right], \tag{2.101}$$

so that, unlike the case for the single hump, R is not constant and decays to zero with time according to eq. (2.101). Eq. (2.100) may now be written as

$$u = \frac{x/t}{1 + e^{x^2/2\delta t}/(e^R - 1)}. \tag{2.102}$$

The solution (2.100) is obviously not self-similar. For $R \ll 1$, the exponential term $(e^R - 1)^{-1}$ in the denominator of eq. (2.102) is much greater than 1 so that u is simply a differentiated Gaussian. This is the stage of the solution where R has decayed sufficiently from its initial large value, the convection has died out, and the flow is essentially diffusive. In the early stages of propagation of the N wave when $R \gg 1$, we may write eq. (2.102) as

$$u \approx \frac{x}{t}\{1 + e^{(x^2/2\delta t - R)}\}^{-1} \tag{2.103}$$

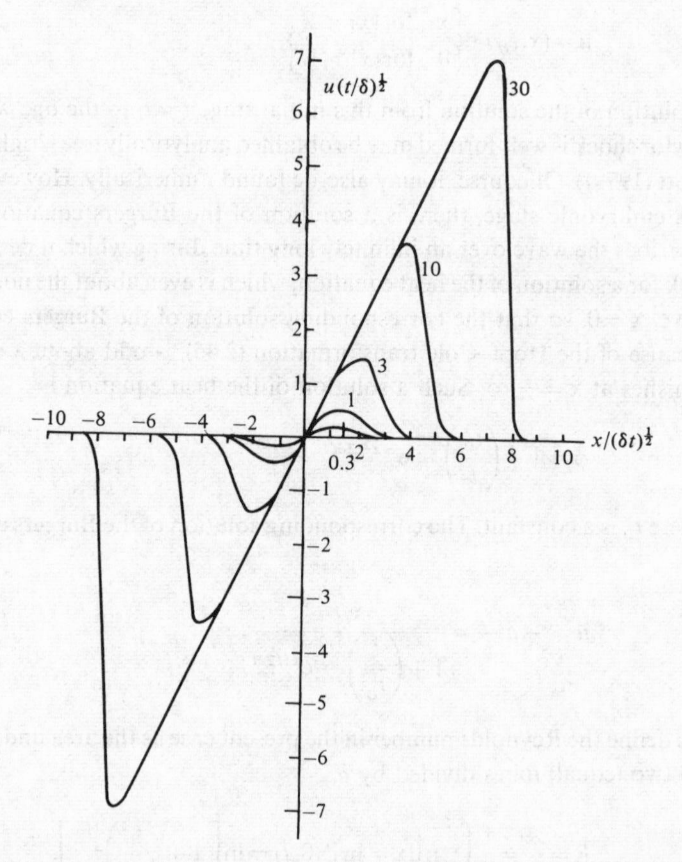

Fig. 2.2. Scaled N wave solutions of Burgers equation for Reynolds numbers 30, 10, 3, 1, 0.3 (from Lighthill (1956)); see sec. 5.6 for evolutionary details.

so that

$$u \sim \begin{cases} \dfrac{x}{t}, & |x| < (2\delta R t)^{1/2}, \\ 0, & |x| > (2\delta R t)^{1/2}. \end{cases} \tag{2.104}$$

This is the inviscid solution of the Burgers equation. The shock centre x_s itself may be found by locating the point such that u is midway between the maximum of the discontinuous profile and zero, that is, when

$$1 + e^{x^2/2\delta t}/(e^R - 1) \approx 2.$$

We thus have

$$x_s = [2\delta t \ln(e^R - 1)]^{1/2} = \left[2\delta t \ln\left(\frac{t_0}{t}\right)^{1/2}\right]^{1/2} \tag{2.105}$$

$$= (\delta \ln t_0)^{1/2} t^{1/2} - \frac{1}{2}\left(\frac{\delta}{\ln t_0}\right)^{1/2} t^{1/2} \ln t.$$

Between the early nearly inviscid phase and the final diffusive phase, there are several other stages, when the shock structure is not given by the Taylor solution and shock wave displacement due to diffusion (see the following paragraph) is not small, and further, when the shock ceases to be thin and becomes of the order of the length of the wave profile itself. The solution is described by the complete form (2.100) (see fig. 2.2). Indeed, much of (singular) perturbation analysis, as we shall see in chapter 3, fails to describe the intermediate stages, beyond the thin shock regime.

The first term in eq. (2.105) is the shock law according to inviscid (weak) shock theory, while the second term is the effect of diffusion and is referred to as shock wave displacement due to diffusion. This is caused in the present case by the mass diffusion across the node of the N wave bringing about its decay (see Lighthill (1956)).

2.8 Periodic initial conditions

Another important solution of the Burgers equation arises from periodic initial conditions. This was first considered by Cole (1951). Subsequently, its counterpart for boundary value problems was obtained by Blackstock (1964) and Parker (1980). If we start with the initial and boundary conditions

$$u(x, 0) = u_0 \sin\frac{\pi x}{l}, \quad 0 \leqslant x \leqslant l, \tag{2.106a}$$

$$u(0,t) = u(l,t) = 0, \quad t > 0, \tag{2.106b}$$

the Hopf–Cole transformation (2.45) gives

$$\begin{aligned}\phi(x,0) &= \exp\left(-\frac{u_0}{\delta}\int_0^x \sin\frac{\pi x}{l}\,\mathrm{d}x\right)\\ &= \exp\left[-\frac{u_0 l}{\delta\pi}\left(1-\cos\frac{\pi x}{l}\right)\right],\end{aligned} \tag{2.107}$$

so that (2.106b) is satisfied. The corresponding solution of the heat equation is

$$\phi(x,t) = A_0 + \sum_{n=1}^{\infty} A_n \exp\left(-\frac{\delta}{2}\frac{n^2\pi^2 t}{l^2}\right)\cos\frac{n\pi x}{l} \tag{2.108}$$

where

$$\begin{aligned}A_0 &= \frac{1}{l}\int_0^l \exp\left[-\frac{u_0 l}{\pi\delta}\left(1-\cos\frac{\pi x}{l}\right)\right]\mathrm{d}x\\ &= \exp\left(-\frac{u_0 l}{\pi\delta}\right)I_0\left(\frac{u_0 l}{\pi\delta}\right),\\ A_n &= \frac{2}{l}\int_0^l \exp\left[-\frac{u_0 l}{\pi\delta}\left(1-\cos\frac{\pi x}{l}\right)\right]\cos\frac{n\pi x}{l}\,\mathrm{d}x\\ &= 2\exp\left(-\frac{u_0 l}{\pi\delta}\right)I_n\left(\frac{u_0 l}{\pi\delta}\right)\end{aligned} \tag{2.109}$$

(see Abramowitz and Stegun (1964)). The solution u of the Burgers equation (2.44) therefore is given by

$$u = \frac{2\delta\pi}{l}\,\frac{\sum_{n=1}^{\infty}\exp(-\delta n^2\pi^2 t/2l^2)\,nI_n\left(\frac{u_0 l}{\pi\delta}\right)\sin\left(\frac{n\pi x}{l}\right)}{I_0\left(\frac{u_0 l}{\pi\delta}\right)+2\sum_{n=1}^{\infty}\exp\left(-\frac{\delta^2 n^2\pi^2 t}{4l^2}\right)I_n\left(\frac{u_0 l}{\pi\delta}\right)\cos\left(\frac{n\pi x}{l}\right)}. \tag{2.110}$$

The value of the denominator at $t = 0$ is

$$I_0(z) + 2\sum_{n=1}^{\infty} I_n(z)\cos\frac{n\pi x}{l} = \exp\left(z\cos\frac{\pi x}{l}\right) \tag{2.111}$$

(see Abramowitz and Stegun (1964)), $z = u_0 l/\pi\delta$, so that u obviously satisfies the initial condition (2.106a) via eq. (2.45). The Reynolds number $R_0 = u_0 l/\delta$ based on the initial amplitude u_0, the length of the initial (periodic) profile l and the viscosity coefficient δ, naturally appears in eq. (2.110). The solution of the heat equation (2.52) subject to the conditions (2.106) is

$$\phi(x,t) = u_0 \exp\left(-\frac{\delta}{2}\frac{\pi^2}{l^2}t\right)\sin\frac{\pi x}{l}. \tag{2.112}$$

A comparison of eqs. (2.110) and (2.112) immediately brings out two aspects of nonlinearity: (1) the generation of an infinity of higher harmonics with diminishing amplitudes, in contrast to the linear solution (2.112) with only the fundamental harmonic, (2) the dependence on the Reynolds number R rather than only on u_0, the initial amplitude of the wave. For $R_0 \to 0$, the functions $I_n(R_0/\pi)$ may be expanded to show that u tends to ϕ with an error of $O(R_0)$. In the limit of large R_0, we may use the asymptotic form of I_n:

$$I_n\left(\frac{R_0}{\pi}\right) \approx \frac{e^{R_0/\pi}}{(2R_0)^{1/2}}\left[1 - \frac{4n^2 - 1^2}{1!\left(\frac{8R_0}{\pi}\right)} + \frac{(4n^2-1^2)(4n^2-3^2)}{2!\left(\frac{8R_0}{\pi}\right)^2} + O\left(\frac{1}{R_0^3}\right)\right]. \tag{2.113}$$

Curiously enough, if only the first term in the expansion (2.113) is used and cancelled throughout in eq. (2.110), the resulting 'approximate' solution

$$u(x,t) \approx \frac{2\delta\pi}{l}\frac{\sum_{n=1}^{\infty}\exp(-\delta n^2\pi^2 t/2l^2) n \sin\frac{n\pi x}{l}}{1 + 2\sum_{n=1}^{\infty}\exp\left(-\frac{\delta n^2\pi^2 t}{2l^2}\right)\cos\frac{n\pi x}{l}} \tag{2.114}$$

turns out to be an exact solution, with its own domain of validity. The exactness of the solution was discovered later by several investigators. A more curious result emerges from expressing (2.114) in yet another form obtained by using theta functions (see Abramowitz and Stegun (1964)). If we use the following results for θ_3,

$$\theta_3(X,T) = 1 + 2\sum_{n=1}^{\infty} e^{-n^2\pi T}\cos 2nX, \tag{2.115}$$

$$X = \frac{\pi x}{2l}, \quad T = \frac{\delta\pi}{2l^2}t,$$

$$\frac{\partial}{\partial X}\ln\theta_3(X,T) = 2\sum_{n=1}^{\infty}(-1)^n\frac{\sin 2nX}{\sinh n\pi T}, \tag{2.116}$$

in eq. (2.114), we have

$$u = -\frac{\delta\pi}{l}\sum_{n=1}^{\infty}\frac{(-1)^n \sin(n\pi x/l)}{\sinh\frac{\delta}{2}(n\pi^2 t/l^2)}. \tag{2.117}$$

Eq. (2.117) gives yet another exact solution! Cole, in his remarkable work, also mentioned that eq. (2.117) approximates to

$$u \approx \frac{l}{t}\left\{\tanh\left(\frac{l-x}{\delta t}\right) - \left(1 - \frac{x}{l}\right)\right\}. \tag{2.118}$$

We shall give later a more rigorous derivation of some of these results for the sinusoidal piston motion treated as a boundary value problem in the light of some recent work, part of which was implicit in Cole's paper. Physically, the sinusoidal initial profile for large R_0 steepens near $x = l$ because of the nonlinear effects manifested in the generation of higher harmonics. This steepening is resisted by the diffusive effects, resulting first in a 'structured' shock near $x = l$. At subsequent times, this effect spreads in the entire profile, leading to exponential decay of the harmonics according to eq. (2.117). There is no 'steep' front at this stage of evolution of the profile. Finally, only the first harmonic survives but with a reduced amplitude as compared to that for a linear wave as given by eq. (2.112). It is interesting to note that the form (2.117) displays no dependence on the initial amplitude.

Reference may be made to Walsh (1969) for the solution of the spatially periodic initial value problem with general initial conditions.

2.9 An equivalent boundary value problem

So far we have been concerned with initial value problems for the Burgers equation and consequently for the heat equation. There is an alternative way of recovering almost all the solutions of Burgers' equation, as listed, for example, by Benton and Platzman (1972), and many more, by suitably posing a boundary value problem, as was shown by Rodin (1970) and Sachdev (1976a). Here we pose such a problem and illustrate it with a few interesting solutions. It is well known (see, for example, Copson (1975)) that the Cauchy–Kowalewsky theorem for the heat equation (2.52), subject to the boundary conditions

$$\phi(0,t) = F(t), \quad \phi_x(0,t) = G(t), \quad t > 0, \quad -\infty < x < \infty, \tag{2.119}$$

gives the following solution provided $F(t)$ and $G(t)$ are analytic in the relevant interval of t:

$$\phi(x,t) = \sum_{n=0}^{\infty} F^{(n)}(t)\left(\frac{2}{\delta}\right)^n \frac{x^{2n}}{(2n)!} + \sum_{n=0}^{\infty} G^{(n)}(t)\left(\frac{2}{\delta}\right)^n \frac{x^{2n+1}}{(2n+1)!}. \tag{2.120}$$

Here the superscript (n) denotes differentiation with respect to t. It turns out that this boundary value problem is equivalent to a pure initial value problem for the heat equation over the same domain as in (2.119). Walsh (1969) has formally proved this equivalence, essentially following Widder (1956).

A corresponding equivalence for the Burgers equation was obtained by way of Hopf–Cole transformation, subject to the condition $\phi(x,0)>0$, by Rodin (1970). From eq. (2.120),

$$\begin{aligned}\phi(x,0) &= \sum_{n=0}^{\infty} F^{(n)}(0)\left(\frac{2}{\delta}\right)^n \frac{x^{2n}}{(2n)!} + \sum_{n=0}^{\infty} G^{(n)}(0)\left(\frac{2}{\delta}\right)^n \frac{x^{2n+1}}{(2n+1)!} \\ &\equiv H(x)\end{aligned} \tag{2.121}$$

for $-\infty < x < \infty$ and eq. (2.120) itself may, in fact, be re-written in the form

$$\phi(x,t) = H(x) + \sum_{n=1}^{\infty} \frac{t^n}{n!} H^{(2n)}(x). \tag{2.122}$$

Now, by way of the Hopf–Cole transformation, the solution of the Burgers equation corresponding to (2.120) is

$$u(x,t) = -\delta \frac{\sum_{n=0}^{\infty} F^{(n+1)}(t)\left(\frac{2}{\delta}\right)^{n+1} \frac{x^{2n+1}}{(2n+1)!} + \sum_{n=0}^{\infty} G^{(n)}(t)\left(\frac{2}{\delta}\right)^n \frac{x^{2n}}{(2n)!}}{\sum_{n=0}^{\infty} F^{(n)}(t)\left(\frac{2}{\delta}\right)^n \frac{x^{2n}}{(2n)!} + \sum_{n=0}^{\infty} G^{(n)}(t)\left(\frac{2}{\delta}\right)^n \frac{x^{2n+1}}{(2n+1)!}}, \tag{2.123}$$

satisfying the boundary conditions

$$u(0,t) = -\delta\frac{G(t)}{F(t)}, \quad u_x(0,t) = \delta\left[\frac{G^2(t)}{F^2(t)} - \frac{2F'(t)}{\delta F(t)}\right]. \tag{2.124}$$

No initial conditions are prescribed. However, this is a well-defined problem and as we have stated earlier is equivalent to an initial value problem over $-\infty < x < \infty$.

Now, we imagine a subsonic piston motion (see Lagerstrom *et al.* (1949)) say, from $x=0$ with displacement $H(t)$ and velocity $H'(t)$. We assume that $H(t)$ is small so that only terms of order $H(t)$ are retained in the Taylor series for $u(H(t),t)$. The boundary condition at the piston is

$$u(H(t),t) = H'(t)$$

or

$$u(0,t) + H(t)u_x(0,t) = H'(t). \tag{2.125}$$

Assuming that $u(0,t)$ is small in comparison with $u_x(0,t)$ and $H'(t)$, we arrive at the following boundary conditions for the Burgers equation:

$$u(0,t)=0,$$
$$u_x(0,t)=\frac{H'(t)}{H(t)}. \tag{2.126}$$

Eqs. (2.124) and (2.126) together imply that

$$\frac{H'(t)}{H(t)}=-2\frac{F'(t)}{F(t)} \tag{2.127}$$

so that, by integration,

$$F(t)=[H(t)]^{-1/2}. \tag{2.128}$$

Moreover, $G(t)=0$. The solution (2.123) now takes the form

$$u(x,t)=-\delta\frac{\sum_{n=0}^{\infty}[(H(t))^{-1/2}]^{(n+1)}\left(\frac{2}{\delta}\right)^{n+1}\frac{x^{2n+1}}{(2n+1)!}}{\sum_{n=0}^{\infty}[H(t)^{-1/2}]^{(n)}\left(\frac{2}{\delta}\right)^{n}\frac{x^{2n}}{(2n)!}}. \tag{2.129}$$

It is evident from eqs. (2.123) and (2.129) that only positive solutions of the heat equation, with precise information about the radius of convergence of their series representations, should be chosen so that the solutions of the Burgers equation remain non-singular. The book by Widder (1975) may be referred to for further information regarding the heat equation.

If we choose the piston displacement as

$$H(t)=\frac{t}{(1+t^{1/2})^2}, \tag{2.130}$$

it is easily verified that (with $\delta=2$ in eq. (2.44) for convenience)

$$u=\frac{x/t}{1+t^{1/2}\,e^{x^2/4t}} \tag{2.131}$$

gives an N wave. The piston motion (2.130) starting from $x=0$ reaches $x=1$ asymptotically as $t\to\infty$.

We now show that the solution (2.117) arising out of a periodic initial condition may, in fact, also be simulated by the 'gentle' piston motion

$$H(t)=\left(1+2\sum_{k=1}^{\infty}e^{-k^2t}\right)^{-2}. \tag{2.132}$$

This function is a strictly increasing function of t such that, like eq. (2.130),

$H(0)=0$ and $H(\infty)=1$. The solution is

$$
\begin{aligned}
u(x,t) &= -\delta\frac{\partial}{\partial x}\ln\left[\sum_{n=0}^{\infty}\left(\frac{2}{\delta}\right)^{n}\left(1+2\sum_{k=1}^{\infty}\mathrm{e}^{-k^2t}\right)^{(n)}\frac{x^{2n}}{(2n)!}\right] \\
&= -\delta\frac{\partial}{\partial x}\ln\left[1+\sum_{n=0}^{\infty}2\left(\frac{2}{\delta}\right)^{n}\left(\sum_{k=1}^{\infty}\mathrm{e}^{-k^2t}\right)^{(n)}\frac{x^{2n}}{(2n)!}\right] \\
&= -\delta\frac{\partial}{\partial x}\ln\left[1+2\sum_{n=0}^{\infty}\left\{\left(\frac{2}{\delta}\right)^{n}\left[\sum_{k=1}^{\infty}(-1)^n k^{2n}\mathrm{e}^{-k^2t}\right]\frac{x^{2n}}{(2n)!}\right\}\right] \\
&= -\delta\frac{\partial}{\partial x}\ln\left[1+2\sum_{k=1}^{\infty}\left\{\mathrm{e}^{-k^2t}\left[\sum_{n=0}^{\infty}(-1)^n\frac{(k\Delta x)^{2n}}{(2n)!}\right]\right\}\right] \\
&= -\delta\frac{\partial}{\partial x}\ln\left[1+2\sum_{k=1}^{\infty}\mathrm{e}^{-k^2t}\cos(k\Delta x)\right] \\
&= -\delta\frac{\partial}{\partial x}\ln\left[\theta_3\left(\frac{\Delta}{2}x,\mathrm{e}^{-t}\right)\right] \\
&= -(2\delta)^{1/2}\sum_{n=1}^{\infty}(-1)^n\frac{\sin(n\Delta x)}{\sinh nt}.
\end{aligned}
\tag{2.133}
$$

In eq. (2.133), Δ denotes $(\delta/2)^{-1/2}$ and the last two steps follow from eqs. (2.115) and (2.116). The rearrangement of terms, term-by-term differentiation, and interchange of the order of summations can be easily justified since the double series and its derivative are absolutely and uniformly convergent.

Our final example is related to (2.118). If we choose

$$u(0,t)=0,\quad u_x(0,t)=\frac{1}{t}\left(1+\frac{\beta^2}{\delta t}\right), \tag{2.134}$$

the conditions for the associated heat equation (2.52) become

$$
\begin{aligned}
&\phi(0,t)=t^{-1/2}\,\mathrm{e}^{\beta^2/2\delta t},\quad \phi_x(0,t)=0,\\
&H(t)=t\,\mathrm{e}^{-\beta^2/\delta t}.
\end{aligned}
\tag{2.135}
$$

Here β is a constant. Substitution of (2.135) into eq. (2.129) leads to

$$u(x,t)=\frac{1}{t}\left(x+\beta\tan\frac{\beta x}{\delta t}\right). \tag{2.136}$$

The solution for $\beta^2>0$ has the form (2.136), while that for $\beta^2<0$ with simple scaling and translation of the variables etc. (see eqs. (2.55)–(2.57)) gives (2.118).

An interesting work on boundary value problems for the Burgers equation (2.44) is due to Kochina (1961), who posed the boundary value problem over a semi-infinite domain,

$$\left.\begin{aligned} &u(0,t)=\psi(t),\\ &\lim_{x\to\infty} u(x,t)=u_\infty\leqslant 0,\\ &0\leqslant x\leqslant\infty,\quad t>0, \end{aligned}\right\} \tag{2.137}$$

requiring the solution to tend to a non-positive value at infinity. The initial function $\psi(t)$ was assumed to have a Fourier series expansion with coefficients of the order of $1/k^r$, $r\geqslant 2$. A periodic solution of the Burgers equation subject to (2.137) was found using the Hopf–Cole transformation, necessitating however the solution of an infinite system of linear algebraic equations. The conditions for the existence of the solution of this system were also analysed.

2.10 Viscosity method and Burgers' equation

The Burgers equation (2.44) may be viewed as a model providing a proper viscous embedding for the inviscid hyperbolic law

$$u_t+uu_x=0. \tag{2.138}$$

More specifically, if u is an 'admissible' solution of eq. (2.138) and u_δ the solution of eq. (2.44) with the same initial conditions, then $u_\delta\to u$ as $\delta\to 0$, in an appropriate sense. It should be noted that eq. (2.138) can be written as a conservation law in more than one way, say,

$$\frac{\partial u}{\partial t}+\frac{\partial}{\partial x}\left(\frac{u^2}{2}\right)=0,\quad \frac{\partial(u^2/2)}{\partial t}+\frac{\partial}{\partial x}\left(\frac{u^3}{3}\right)=0,\quad \text{etc.} \tag{2.139}$$

and there can be many (in fact an infinity of) possible Rankine–Hugoniot conditions arising from these (infinite) conservation laws (see Courant and Hilbert (1962)). The embedding of eq. (2.139) with viscosity facilitates selection of just the correct conservation law, satisfying the so-called entropy condition (see Gelfand (1959), Lax (1957) and Dafermos (1974)). Hopf (1950) discussed the limiting behaviour of eq. (2.44) as $\delta\to 0$.

In our context, however, we regard the Burgers equation and its generalisations as models arising out of actual physical conditions and not merely serving to sift a unique solution for a given hyperbolic equation. However, the relation between the viscous and inviscid solutions is very important. Indeed the inviscid solution always forms an important component of the solution, holding in some domain. For example, for both the single hump and the N wave, the solution $u = x/t$ holds in the essentially inviscid domain to the fore of the shock layer when high Reynolds number conditions prevail. As we have mentioned before, Whitham (1974), using the method of steepest descent for the integrals in eq. (2.73), verified that, in the limit of $\delta \to 0$, the solutions of the Burgers equation approach those of eq. (2.138) satisfying the Rankine–Hugoniot relation across the shock discontinuity, namely

$$U = \tfrac{1}{2}(u_1 + u_2), \tag{2.140}$$

where U is the velocity of the shock and u_1, u_2 are values of u immediately ahead of and behind the shock, respectively. In subsequent chapters, particularly when perturbation methods are discussed to solve generalised Burgers equations, the inviscid solution will again form the outer solution and will be matched to the inner solution in the shock layer.

3 Generalised Burgers equations

3.1 Introduction

The Burgers equation (2.44) is an idealised equation which combines a simple nonlinearity with a small linear viscous term. In actual physical situations there are other complicating physical factors which alter this equation, resulting, in general, in the loss of a Hopf–Cole type transformation for exact linearisation. These contributing terms may be lower order source or sink terms or geometrical expansion terms so that we may have equations, for example, of the type

$$u_t + uu_x + \lambda u = \frac{\delta}{2}u_{xx}, \tag{3.1}$$

$$u_t + uu_x + \frac{ju}{2t} = \frac{\delta}{2}u_{xx}, \quad j = 1, 2. \tag{3.2}$$

In eq. (3.1) if the coefficient λ is positive, we have a sink term which, in the absence of viscosity in the early stages of the wave ($\delta = 0$), will dampen the wave so much that no shock is formed. On the other hand, a negative λ will accelerate the formation of the shock. Eq. (3.2) combines the effect of spherical or cylindrical expansion ($j = 2, 1$, respectively) with nonlinearity and diffusion, and arises from the corresponding geometrical shape of the source. This, too, has the effect of dampening the amplitude of the wave. The amplitude reduction due to these lower order 'friction' terms may be contrasted with that due to the diffusive damping represented by the higher order viscous term $\frac{1}{2}\delta u_{xx}$. Here the wave is caused to spread or diffuse, and dampen. There may be other higher order effects, say, due to dispersion, leading to an equation of the type

$$u_t + uu_x = \frac{\delta}{2}u_{xx} - \mu u_{xxx}, \quad \mu > 0, \tag{3.3a}$$

the so-called Korteweg–de Vries–Burgers equation. This would admit a predominantly shock-like structure with small oscillations in the tail of the shock if $\mu \ll \delta$. Conversely, it may describe essentially dispersive phenomena such as in tidal waves, where account is taken of the small 'eddy viscosity' which may be present. This would happen if $\mu \gg \delta$. A nice discussion of the solutions resulting from different sizes of these terms may be found in Johnson (1970). We shall have occasion to refer again to this equation and its stationary propagating solutions later when we discuss Fisher's equation which, in a strange way, has some connection with eq. (3.3a).

These (model) generalised Burgers equations (see Crighton (1979) for further examples) can be derived from the basic system of equations using, for example, the method of multiple scales. We shall illustrate this by deriving eq. (3.2) in sec. 3.3.

In fact, as we remarked earlier, the following sequence of events takes place when a forcing agency, a fast piston, for example, pushes the gas and stops. A pulse is created which due to nonlinearity may steepen at its front. It may be aided or retarded by a lower order term in acquiring a discontinuity, a shock, at its head; in the absence of such a term, the case $\lambda = 0$, an initial profile however smooth will necessarily steepen into a shock. Until the formation of the shock, the wave is governed by eq. (3.1) with $\delta = 0$. Viscous diffusion comes into play to 'loosen' the front. It takes some time for the nonlinear and viscous effects to come to a certain 'stationary' balance, when we have a thin Taylor shock. This phase of the evolution of the wave persists for some time until the shock has thickened to become a sizeable fraction of its total length. Then, in general, the stationary Taylor structure is no longer valid. The pulse continues to broaden and dampen under the influence of diffusive effects until its amplitude has diminished so much that the nonlinear term becomes negligible and the wave evolves under pure diffusion (and lower order damping), dying out after a very long time.

Mathematically, before the formation of the shock, the flow is governed by the inviscid form of eq. (3.1),

$$u_t + uu_x + \lambda u = 0. \tag{3.3b}$$

Indeed the solution given by this equation holds for a long time if one ignores the thickness of the shock so that it may be treated as a sharp discontinuity. We may thereafter switch to eq. (3.1) as representing a singular perturbation problem with a small coefficient δ multiplying the highest order derivative. We obtain the solution as a matched asymptotic expansion. This solution holds for some time after which, in general, it becomes invalid in some domains. The exception to this is the Burgers equation itself, for which, in the case of a sinusoidal piston motion, a first order (matched) composite solution turns out to be an exact solution valid for all time. (This particular solution, however, has its own domain of validity). Actually, the periodicity here is imposed in the boundary condition, and space and time co-ordinates interchange their roles (Parker, 1980). Generally, however, the first order matched asymptotic solution ceases to be valid because (i) the shock is no longer thin compared to the overall pulse length, (ii) the shock displacement due to diffusion takes the shock too far from its location according to inviscid or weak shock theory, (iii) the solution acquires a form different from the Taylor shock (see Crighton and Scott (1979)).

When any of the above conditions comes about, it is difficult to find the solution analytically until the pulse has died out sufficiently so as to evolve according to the linearised form of the equation. Even then the solution is unknown to the extent of a constant multiple which is a remnant of and a link with its earlier evolution. Another notable exception is a physically relevant and correct similarity solution, when one exists, to which a certain class of initial pulses evolves. Thus, the similarity solution takes over as an intermediate asymptotic – a nonlinear solution persisting all the way to infinity and coinciding with the linear solution in the appropriate final regime.

In this chapter, we shall discuss singular perturbation approaches as applied to several different physical problems, modelled by Burgers' equation and its generalised forms. We shall study both initial and boundary value problems. We shall also derive the non-planar Burgers equation for spherical and cylindrical geometries by the method of multiple scales, and solve it by the method of matched asymptotic expansion and an alternative analytic approach which uses an infinite sum of the products of functions of similarity and time variables. We study GBEs and their transformations; the inhomogeneous Burgers equation is considered in some detail, applying it, in particular, to acoustic waves excited by the absorption of laser radiation.

Remarks on singular perturbation techniques

A quick glance at the Burgers equation (2.44) reveals that the small parameter δ, the coefficient of viscous diffusion, multiplies the highest order derivative in the equation, suggesting a treatment by what are referred to as singular perturbation techniques. A power series solution of the problem in δ is called regular, if it can be expanded in the form

$$u(x,t) = u_0(x,t) + \delta u_1(x,t) + \delta^2 u_2(x,t) + \cdots,$$

so that this series in δ has a non-vanishing radius of convergence. The exact solution for small but non-zero δ smoothly approaches the unperturbed or zeroth order solution $u_0(x, t)$ as $\delta \to 0$. In contrast, a singular perturbation problem may not have a power series solution at all.

A hallmark of the singular nature of the equation is, as we noted earlier, the multiplication of the highest order derivative term by a small parameter so that the unperturbed form (the case with $\delta = 0$) of the equation has its order depressed by 1, thereby affecting its ability to satisfy all the initial and boundary conditions relevant to the problem. If this unperturbed equation has a solution, its nature is fundamentally different from the exact solution, particularly in some (thin) layers referred to as boundary layers or shock layers, which are characterised by relatively sharp changes in the solution.

Such nonlinear equations, in general, cannot be solved in a closed form. The perturbation methods break the problem into several parts, for each of which the solution has either a regular perturbation series or a singular perturbation series holding in some sub-domain. Then, these separate solutions are matched to yield a global approximation of the solution, the so-called uniform approximation. The method is to find the outer solution (of the 'outer limit' of the equation as $\delta \to 0$) in the region outside the shock layer as a regular perturbation series. Similarly, an inner solution (of the 'inner limit' of the equation as $\delta \to +0$) is found which holds in the shock layer. This is accomplished by introducing an inner variable with respect to which the variation of the dependent variable is not rapid. An order-by-order matching over an overlapping region is carried out by introducing the intermediate limit such that the outer independent variable tends to zero, while the inner variable tends to infinity; the existence of an overlap region is normally assumed. The extent of the overlap region, while finite in terms of the outer variable, is infinite in terms of the inner variable. These overall ideas will become clear as we take up singular perturbation analysis of several generalised Burgers equations. There are now available several books on perturbation methods, to which reference may be made for further details (Nayfeh, 1973, Van Dyke, 1975, Kevorkian & Cole, 1981).

It is well to recall relevant features of the exact single hump and N wave solutions of the standard Burgers equation (see secs. 2.6 and 2.7) to motivate the perturbation approach to the solution for the generalised Burgers equations, for which exact solutions cannot be found; these solutions however have a qualitatively similar structure. For the single hump case, in the limit of $\delta \to 0$ (that is, Reynolds number tending to infinity) the solution is given by a straight ramp (2.94) in $0 < x < (2At)^{1/2}$ and 0 outside. This is an exact solution of the outer limit of eq. (2.44), namely $u_t + uu_x = 0$, which we have earlier termed an inviscid Burgers equation. The shock in this case is located at $x = (2At)^{1/2}$ and has a velocity $U = (A/2t)^{1/2}$ (see (2.88) for the definition of A). The shock condition obtained from the conservation law $u_t + (\partial/\partial x)(u^2) = 0$, namely, $U = \frac{1}{2}(2A/t)^{1/2} = (A/2t)^{1/2}$, is automatically satisfied. This is also referred to as weak shock theory. When δ is small but finite, so that the Reynolds number is large and finite, the shock centred at $x = (2At)^{1/2}$ has a small thickness, and a structure, which, to first order in δ, agrees with that given by Taylor's shock theory (see sec. 2.5). Similar discussion applies to the profile of the N wave. In particular, the shock displacement due to diffusion is given by the second term on the right of eq. (2.105). We shall seek such a solution using singular perturbation methods, at least to first order in δ, for generalised Burgers equations; in principle, even higher order solutions may be found but, in practice, they turn out to be highly intricate (see Lardner (1986)).

3.2 Generalised Burgers equation with damping

We consider eq. (3.1) with a slight change so that it coincides with the form considered by Lardner and Arya (1980). This equation arises from considering the plane motion of a continuous medium for which the constitutive relation for the stress contains a large linear term proportional to the strain, a small term which is quadratic in the strain, and a small dissipative term proportional to the strain-rate (Lardner, 1976). We make the transformation

$$u \to h, \quad x \to -\theta, \tag{3.4}$$

so that eq. (3.1) changes into

$$h_t - hh_\theta + \lambda h = \frac{\delta}{2} h_{\theta\theta}, \quad \delta \ll 1, \quad \lambda > 0, \tag{3.5}$$

with the initial condition

$$h(\theta, 0) = H(\theta)\,(\text{say}). \tag{3.6}$$

The term λh represents a small viscous damping, proportional to velocity. We first obtain the outer (regular) perturbation solution

$$h = h^{(0)} + \delta h^{(1)} + \delta^2 h^{(2)} + \cdots. \tag{3.7}$$

Substitution of (3.7) into eq. (3.5) shows that the lowest order term $h^{(0)}$ satisfies the equation

$$h_t^{(0)} - h^{(0)} h_\theta^{(0)} + \lambda h^{(0)} = 0, \tag{3.8}$$

whose characteristic form

$$\frac{\mathrm{d}t}{1} = \frac{\mathrm{d}\theta}{-h^{(0)}} = \frac{\mathrm{d}h^{(0)}}{-\lambda h^{(0)}} \tag{3.9}$$

gives the explicit solution

$$h^{(0)}(\theta, t) = \mathrm{e}^{-\lambda t} H(\theta_1), \tag{3.10}$$

$$\theta_1 = \theta + \sigma H(\theta_1), \tag{3.11}$$

$$\sigma = \frac{1}{\lambda}(1 - \mathrm{e}^{-\lambda t}). \tag{3.12}$$

At $t = 0$ (i.e., $\sigma = 0$), $\theta = \theta_1$, and $h^{(0)}$ satisfies the initial condition (3.6). It is obvious from eq. (3.6) that the higher order terms $h^{(1)}$ etc. are to be found subject to zero initial conditions. θ_1 is the characteristic variable so that, for $\theta_1 =$ constant, eq. (3.11) describes the characteristics in the θ–σ plane, σ replacing t as a convenient (modified) time variable. These characteristics are straight lines carrying different initial values of $H(\theta_1)$ corresponding to different values of θ_1. It is interesting to note that the characteristics will meet and form an envelope if

$$1 = \sigma H'(\theta_1), \tag{3.13}$$

obtained by differentiating eq. (3.11) with respect to θ_1. The earliest time this happens is given by

$$\sigma_{\mathrm{f}} = [H'(\theta_{\mathrm{f}})]^{-1}, \tag{3.14}$$

where θ_{f} is the value of θ_1 at which $H'(\theta_1)$ is maximum. Until $\sigma = \sigma_{\mathrm{f}}$, the solution is smooth and is given by eqs (3.10)–(3.12). Indeed, if $\lambda^{-1} < \sigma_{\mathrm{f}}$, the corresponding smooth solution will remain valid right up to $t = \infty$, as follows easily from eq. (3.12), and the viscous term does not come into play at all. However, in general, characteristics will meet for $\sigma > \sigma_{\mathrm{f}}$ and form a shock. Condition (3.14) requires a quite steep initial profile for an early formation of the shock.

It is convenient to introduce the characteristic variable θ_1, and σ, to transform the first order equation

$$h_t^{(1)} - \mathrm{e}^{-\lambda t} H(\theta_1) h_\theta^{(1)} + \lambda h^{(1)} - \mathrm{e}^{-\lambda t} H'(\theta_1)[1 - \sigma H'(\theta_1)]^{-1} h^{(1)}$$
$$= \tfrac{1}{2}\mathrm{e}^{-\lambda t} H''(\theta_1)[1 - \sigma H'(\theta_1)]^{-3} \tag{3.15}$$

(which follows from eq. (3.5) and (3.7) by equating terms of order δ) into

$$\frac{\partial}{\partial \sigma}\{[1 - \sigma H'(\theta_1)] g(\theta_1, \sigma)\}$$
$$= \tfrac{1}{2} H''(\theta_1)[1 - \sigma H'(\theta_1)]^{-2} (1 - \lambda \sigma)^{-1}, \tag{3.16}$$

where

$$g(\theta_1, \sigma) = e^{\lambda t} h^{(1)}.$$

The right hand side of eq. (3.16) can easily be put into partial fractions, and an integration with respect to σ together with the condition $h^{(1)}(\theta, 0) = 0$ determines $h^{(1)}$. The outer solution of eq. (3.5), to first order in δ, can thus be written as

$$\begin{aligned} h &= e^{-\lambda t} H(\theta_1) \\ &\quad + \tfrac{1}{2}\delta H''(\theta_1) e^{-\lambda t} \left\{ \frac{\sigma H'(\theta_1)}{[H'(\theta_1) - \lambda][1 - \sigma H'(\theta_1)]^2} \right. \\ &\quad \left. + \frac{\lambda}{[H'(\theta_1) - \lambda]^2 [1 - \sigma H'(\theta_1)]} \ln\left[\frac{1 - \sigma H'(\theta_1)}{1 - \lambda\sigma}\right] \right\}. \end{aligned} \tag{3.17}$$

To get the inner solution, we introduce the stretched inner variable $\alpha = \delta^{-1}[\theta - \theta_s(t)]$, where $\theta = \theta_s(t)$ is the trajectory of the shock wave to be determined. Eq. (3.5), in terms of the variables α and t, becomes

$$\frac{1}{2}\frac{\partial^2 h}{\partial \alpha^2} + [h + \theta_s'(t)]\frac{\partial h}{\partial \alpha} = \delta \frac{\partial h}{\partial t} + \delta \lambda h. \tag{3.18}$$

The solution of this equation, the inner solution, may be sought in the form

$$h(\alpha, t) = h^{(0)}(\alpha, t) + \delta h^{(1)}(\alpha, t) + \cdots. \tag{3.19}$$

The dependence on the arguments will distinguish this expansion from (3.7). The equation for $h^{(0)}$ becomes an 'ordinary' differential equation

$$\frac{1}{2}\frac{\partial^2 h^{(0)}}{\partial \alpha^2} + [h^{(0)} + \theta_s'(t)]\frac{\partial h^{(0)}}{\partial \alpha} = 0 \tag{3.20}$$

whose solution can easily be written as

$$h^{(0)}(\alpha, t) = -\theta_s'(t) + a(t)\tanh\{a(t)[\alpha + b(t)]\}, \tag{3.21}$$

where $a(t)$ and $b(t)$ are 'constants' of integration. The equation for $h^{(1)}(\alpha, t)$, from eq. (3.18) and (3.19), is

$$\begin{aligned} \frac{1}{2}\frac{\partial^2 h^{(1)}}{\partial \alpha^2} + [h^{(0)} + \theta_s'(t)]\frac{\partial h^{(1)}}{\partial \alpha} + \frac{\partial h^{(0)}}{\partial \alpha} h^{(1)} &= -\theta_s^*(t) \\ &\quad + e^{-\lambda t}\frac{\partial}{\partial t}[e^{\lambda t} a \tanh a(\alpha + b)], \end{aligned} \tag{3.22}$$

$$\theta_s^* = e^{-\lambda t}\frac{\partial}{\partial t}[e^{\lambda t}\theta_s'(t)],$$

wherein eq. (3.21) has been used to bring about some simplification. Eq. (3.22) is a linear nonhomogeneous 'ordinary' differential equation in $h^{(1)}$ satisfying zero initial conditions. A particular integral, after substitution for $h^{(0)}$, and some effort, leads to the inner solution to $O(\delta)$:

$$\begin{aligned} h = &-\theta_s'(t) + aT - \delta\theta_s^*\left[\tfrac{1}{2}\alpha^2(1-T^2) + \frac{\alpha}{a}T - \frac{1}{4a^2}(1+T^2)\right] \\ &+ \delta c\left[\alpha(1-T^2) + \frac{T}{a}\right] \\ &+ \delta\left[\frac{a'\alpha}{2a^2}(1+T^2) - \frac{a'}{2a^2}T + \frac{(ab)'}{2a}(1+T^2) + d(1-T^2)\right] \\ &+ \frac{\delta\lambda}{a}\left[T\ln\cosh a(\alpha+b) + \tfrac{1}{2}a(\alpha+b)(1-T^2)\right. \\ &\left. - \tfrac{1}{2}T + (1-T^2)\int^{a(\alpha+b)} \ln\cosh u\, du\right], \end{aligned} \tag{3.23}$$

where $T = \tanh a(\alpha + b)$, and $c(t)$ and $d(t)$ are additional 'constants' of integration. The prime denotes differentiation with respect to t. To match eq. (3.23) with the outer solution (3.17), we find its intermediate limits as $\alpha = [\theta - \theta_s(t)]/\delta$ tends to $\pm\infty$.

As $\alpha \to +\infty$, $T \to 1$, and eq. (3.23) becomes

$$\begin{aligned} h \sim &-\theta_s'(t) + a + \delta\alpha[\lambda + a^{-1}(a' - \theta_s^*)] \\ &+ \delta a^{-1}[(ab)' + c + (2a)^{-1}(\theta_s^* - a') \\ &+ \lambda ab - \lambda(\tfrac{1}{2} + \ln 2)]. \end{aligned} \tag{3.24}$$

The terms $(\delta\lambda T/a)\ln\cosh a(\alpha + b)$ in eq. (3.23) contributes $\delta\alpha\lambda, \delta\lambda b$ and $-\delta\lambda a^{-1}\ln 2$; other limits are easily obtained. Exponentially small terms have been ignored.

Similarly, as $\alpha \to -\infty$, $T \to -1$, and eq. (3.23) leads to

$$\begin{aligned} h \sim &-\theta_s'(t) - a + \delta\alpha[\lambda + a^{-1}(a' + \theta_s^*)] \\ &+ \delta a^{-1}[(ab)' - c + (2a)^{-1}(\theta_s^* + a') + \lambda ab + \lambda(\tfrac{1}{2} + \ln 2)]. \end{aligned} \tag{3.25}$$

To obtain the shock locus, $\theta = \theta_s(t)$, we find the characteristics meeting it at each point t from the right and from the left, according to weak shock theory. These are denoted respectively by $\theta = \theta_1^+$ and $\theta = \theta_1^-$. A first approximation to these is obtained by using the lowest order outer solution (3.11),

$$\theta_1^{\pm} = \theta_s + \sigma H(\theta_1^{\pm}). \tag{3.26}$$

Substracting eq. (3.26) from eq. (3.11), we have

$$\begin{aligned}\theta_1 - \theta_1^{\pm} &= \theta - \theta_s + \sigma[H(\theta_1) - H(\theta_1^{\pm})] \\ &\approx \theta - \theta_s + \sigma H'(\theta_1^{\pm})(\theta_1 - \theta_1^{\pm})\end{aligned} \tag{3.27}$$

in a close neighbourhood of the shock $\theta \approx \theta_s$. Eq. (3.27) can be rewritten as

$$(\theta_1 - \theta_1^{\pm}) \approx (\theta - \theta_s)[1 - \sigma H'(\theta_1^{\pm})]^{-1}, \tag{3.28}$$

giving the approximate characteristics on the two sides of the shock.

In the vicinity of the shock locus, the term $H(\theta_1)$ in the outer solution (3.17) can be written as

$$H(\theta_1) \approx H(\theta_1^{\pm}) + H'(\theta_1^{\pm})[1 - \sigma H'(\theta_1^{\pm})]^{-1}(\theta - \theta_s), \tag{3.29}$$

using eq. (3.28). Thus the inner approximation of the outer solution (3.17) is

$$h(\theta, t) \sim e^{-\lambda t}\{H(\theta_1^{\pm}) + H'(\theta_1^{\pm})[1 - \sigma H'(\theta_1^{\pm})]^{-1}(\theta - \theta_s)\} + \delta K^{\pm} \quad \text{as } \theta \to \theta_s \pm 0, \tag{3.30}$$

where

$$K^{\pm} = \tfrac{1}{2}H''(\theta_1^{\pm})e^{-\lambda t}\left\{\frac{\sigma H'(\theta_1^{\pm})}{[H'(\theta_1^{\pm}) - \lambda][1 - \sigma H'(\theta_1^{\pm})]^2} + \frac{\lambda}{[H'(\theta_1^{\pm}) - \lambda]^2[1 - \sigma H'(\theta_1^{\pm})]}\ln\left[\frac{1 - \sigma H'(\theta_1^{\pm})}{1 - \lambda\sigma}\right]\right\}. \tag{3.31}$$

To match (3.30)–(3.31) to the inner (shock layer) solution (3.23) we substitute $\alpha = (\theta - \theta_s)/\delta$ in the outer limits (3.24)–(3.25) of the inner solution. We get the following six matching conditions,

$$-\theta_s' \pm a = e^{-\lambda t}H(\theta_1^{\pm}), \tag{3.32}$$

$$\lambda + a^{-1}(a' \mp \theta_s^*) = e^{-\lambda t}H'(\theta_1^{\pm})[1 - \sigma H'(\theta_1^{\pm})]^{-1}, \tag{3.33}$$

$$a^{-1}[(ab)' \pm c + (2a)^{-1}(\theta_s^* \mp a') + \lambda ab \mp \lambda(\tfrac{1}{2} + \ln 2)] = K^{\pm}, \tag{3.34}$$

resulting from equating terms of order zero, $(\theta - \theta_s)$, and δ, respectively. The first pair of conditions (3.32) gives expressions for the zero order jump in h across the shock, $2a$, and for the shock velocity, namely

$$2a = e^{-\lambda t}[H(\theta_1^+) - H(\theta_1^-)], \tag{3.35}$$

$$\theta_s'(t) = -\tfrac{1}{2}e^{-\lambda t}[H(\theta_1^+) + H(\theta_1^-)]. \tag{3.36}$$

Eqs. (3.36) and (3.26) together give three relations among the three unknowns $\theta_1^{\pm}$ and $\theta_s(t)$, the functional form of H being known from the initial condition. Then eq. (3.35) gives the function a. For $\lambda = 0$, eq. (3.36) simply expresses the well-known geometrical rule for the shock direction at any point as the mean of the directions of the two characteristics meeting the point from both sides of the shock. Eqs. (3.33) do not give any new information and can, in fact, be obtained by differentiating eqs. (3.32) with respect to t and using eqs. (3.35)–(3.36). The pair of eqs. (3.34) provides the first order effects due to viscosity. Adding these equations we get

$$\frac{\mathrm{d}}{\mathrm{d}t}(ab) + \lambda(ab) = -\frac{\theta_s^*}{2a} + \left(\frac{a}{2}\right)(K^+ + K^-) \tag{3.37}$$

where K^+ and K^- are given by eq. (3.31) so that

$$\begin{aligned}\frac{\mathrm{d}(ab)}{\mathrm{d}t} + \lambda(ab) = \tfrac{1}{4}\mathrm{e}^{-\lambda t}\frac{\mathrm{d}}{\mathrm{d}t}\Bigg\{ &\ln\left[\frac{1-\sigma H'(\theta_1^-)}{1-\sigma H'(\theta_1^+)}\right] \\ &+ \frac{\lambda}{H'(\theta_1^-)-\lambda}\ln\left[\frac{1-\sigma H'(\theta_1^-)}{1-\lambda\sigma}\right] \\ &- \frac{\lambda}{H'(\theta_1^+)-\lambda}\ln\left[\frac{1-\sigma H'(\theta_1^+)}{1-\lambda\sigma}\right]\Bigg\}.\end{aligned} \tag{3.38}$$

This equation can be integrated for ab. Substituting for $a(t)$ from eq. (3.35), we thus have

$$\begin{aligned}b(t) = \frac{1}{2[H(\theta_1^+) - H(\theta_1^-)]}\Bigg\{ &\ln\left[\frac{1-\sigma H'(\theta_1^-)}{1-\sigma H'(\theta_1^+)}\right] \\ &+ \frac{\lambda}{H'(\theta_1^-)-\lambda}\ln\left[\frac{1-\sigma H'(\theta_1^-)}{1-\lambda\sigma}\right] \\ &- \frac{\lambda}{H'(\theta_1^+)-\lambda}\ln\left[\frac{1-\sigma H'(\theta_1^+)}{1-\lambda\sigma}\right]\Bigg\}.\end{aligned} \tag{3.39}$$

In the derivation of eq. (3.39), we have determined the constants of integration by imposing the condition that $\theta_1^- = \theta_1^+$ at the inception of the shock at $t = t_f$, and hence that $a = 0$ there. The zeroth order solution (3.21) clearly shows that the centre of the (structured) tanh shock is at $b(t) = -\alpha$(that is, at $\theta = \theta_s - \delta b(t)$) so that $b(t)$ provides the important quantity we have earlier referred to as the shock displacement due to diffusion. According to eqs. (3.10) and (3.32), $2a$ is the zeroth order jump in h across the shock, while (by subtracting the conditions in the pair (3.34))

$$\delta a^{-1}(2c - \lambda - 2\lambda \ln 2) = \delta(K^+ - K^-) \tag{3.40}$$

gives the first order correction to this jump (cf. the zero order solution

(3.21)). The formula (3.39) gives $-\delta b$, the shock wave displacement due to diffusion, in agreement with Lighthill's (1956) expression (see eqs. (140) and (159) of his paper) for the standard Burgers equation, if we put $\lambda = 0$ and $H'(\theta_1^+) = 0$.

Lardner and Arya (1980) have also given a similar perturbation analysis for the more general equation

$$\frac{\partial h}{\partial t} = [\mu h + \nu h^2 + \nu c(t)]\frac{\partial h}{\partial \theta} + \tfrac{1}{2}\delta\frac{\partial^2 h}{\partial \theta^2} \tag{3.41}$$

which arises in the consideration of the motions of a continuous medium when the stress–strain relation contains a term cubic in the strain in addition to the terms included in eq. (3.5).

A related study is due to Murray (1970a), who considered the model equation

$$u_t + (u + a)u_x + \lambda u = 0, \quad a > 0, \quad \lambda > 0, \tag{3.42}$$

and an initial profile of a finite length $0 \leqslant x \leqslant X$ with a shock at its front, $x = X$. Eq. (3.42) differs from eq. (3.3) due to the presence of the constant a in the convective term which makes a non-trivial change in the nature of the solution. In fact, the behaviour of the solution also depends on whether $2\lambda X > 1$ or $2\lambda X < 1$. For large time, the solution always decays exponentially like $u(x, t) = O(e^{-\lambda t})$ in a finite distance if $a = 0$, and in an infinite distance if $a \neq 0$. Murray has also considered the more general equation

$$u_t + g(u)u_x + \lambda h(u) = 0, \quad \lambda > 0, \quad g_u(u) > 0, \quad h_u(u) > 0 \quad \text{for } u > 0,$$

where

$$h(u) = O(u^\alpha), \quad \alpha > 0, \quad \text{and} \quad 0 < u \ll 1. \tag{3.43}$$

The asymptotic form of propagation and decay of a single hump pulse with a shock at its head depends crucially on the exponent α. Under the assumptions in eq. (3.43), the disturbance decays (i) within a finite time and finite distance for $0 < \alpha < 1$ and is described by a unique solution under certain conditions, (ii) within an infinite time, like $O(e^{-\lambda t})$, and in a finite distance for $\alpha = 1$, (iii) within an infinite time and distance like $O(t^{-1/(\alpha-1)})$ for $1 < \alpha \leqslant 3$, and (iv) like $O(t^{-1/2})$ for $\alpha \geqslant 3$. Thus, $\alpha = 1$ and $\alpha = 3$ appear to have special significance as points of bifurcation. Eq. (3.42) arises from the consideration of stress wave propagation in a mildly nonlinear Maxwell rod with a finite nonlinear viscous damping. Murray has also given the first order correction to the asymptotic solution when (small) viscous diffusion is included, and $\lambda = 0$ (see also Murray (1970b)).

In sec. 5.7, we shall give a similarity analysis of the viscous analogue of eq. (3.43), that is, the generalised Burgers equation

$$u_t + u^\beta u_x + \lambda u^\alpha = \frac{\delta}{2} u_{xx}, \tag{3.44}$$

where α and β are positive parameters, and bring out the relation with Murray's asymptotic study, in particular the dependence of the solution on the parameter α. These results will also be verified by comparison with the numerical solution.

3.3 Derivation of the non-planar Burgers equation

When waves are produced by strong impulses or large explosions, they are mainly governed by these forcing agencies for a relatively short period. Thereafter, they propagate to infinite distance and for infinite time under the influence of effects such as small nonlinearity, small dissipation and small geometrical damping. Thus, although the wave propagation is not far from linear, these small effects, individually or in some combination, accumulate and any regular perturbation scheme gives a description of the wave which is valid to a finite distance and time. This necessitates a perturbation scheme which enlarges these domains of validity and which may deliver in the process a simpler (model) equation as a descriptor of the wave for, possibly, an infinite distance and time. This is what is accomplished by the so-called method of multiple scales. It takes into account the presence of several time (or distance) scales over which small effects gather up to produce a cumulative effect. These times, referred to as slow times, are formally introduced as new independent variables, in addition to the standard time, namely the fast time. Thus, the number of independent variables in the differential equation increases, and the generality so provided is exploited so that the non-uniformity, produced by the small nonlinear effects, say, in the form of 'secular' terms proportional to time in a regular perturbation scheme, is eliminated and the domain of validity of the perturbation scheme enlarged. Although these ideas are best illustrated by way of application to ordinary differential equations (see Kevorkian and Cole (1981)), we here refer to an example from gas dynamics, explaining the difficulty that arises from a single small effect, namely nonlinearity, when the motion is produced by a slow piston. We shall later on deal with the three effects simultaneously, namely nonlinearity, geometrical spreading and viscous diffusion.

We consider an ambient medium at rest with uniform density ρ_0 and pressure p_0, and envisage a piston motion with a characteristic time T_0 such that the distance traversed by the piston is small as compared to $A_0 T_0$ where A_0 is equal to the ambient sound speed $(\gamma p_0/\rho_0)^{1/2}$. The piston motion therefore may be written as

$$X_p(T) = \varepsilon A_0 T_0 f\left(\frac{T}{T_0}\right), \quad \varepsilon \ll 1, \quad \dot{X}_p(T) = \varepsilon A_0 f'\left(\frac{T}{T_0}\right), \tag{3.45}$$

We denote the dimensional quantities by capital letters and corresponding non-dimensional ones by small letters; we thus have

$$u(x,t) = \frac{U(X,T)}{A_0}, \quad a(x,t) = \frac{A(X,T)}{A_0}, \quad x = \frac{X}{A_0 T_0}, \quad t = \frac{T}{T_0}, \tag{3.46}$$

where $u(x,t)$ and $a(x,t)$ are the particle velocity and sound speed, respectively.

Since the piston motion is small, the shocks produced will be weak so that we may assume isentropic conditions. The mathematical problem may thus be stated as follows (see eqs. (2.10)–(2.11)):

$$\frac{\partial a}{\partial t} + u\frac{\partial a}{\partial x} + \frac{\gamma-1}{2} a \frac{\partial u}{\partial x} = 0, \tag{3.47}$$

$$\frac{\partial u}{\partial t} + u\frac{\partial u}{\partial x} + \frac{2}{\gamma-1} a \frac{\partial a}{\partial x} = 0, \tag{3.48}$$

with initial conditions

$$t = 0, \quad x > 0, \quad u = 0, \quad a = 1. \tag{3.49}$$

The boundary condition on the piston $x_p = \varepsilon f(t)$ gives the particle velocity there as

$$u(\varepsilon f(t), t) = \frac{\mathrm{d}}{\mathrm{d}t} x_p = \varepsilon f'(t), \quad t > 0. \tag{3.50}$$

A straightforward perturbation scheme has

$$\left.\begin{aligned} u(x,t) &= \varepsilon u_1(x,t) + \varepsilon^2 u_2(x,t) + \cdots, \\ a(x,t) &= 1 + \varepsilon a_1(x,t) + \varepsilon^2 a_2(x,t) + \cdots, \end{aligned}\right\} \tag{3.51}$$

$$u(0,t) + \varepsilon f(t) u_x(0,t) + O(\varepsilon^2) = \varepsilon f'(t), \tag{3.52}$$

eq. (3.52) following from eq. (3.50). Substitution of (3.51) into eqs. (3.47)–(3.49) and (3.52) gives the following system of equations for first and second order approximations:

$$\frac{\partial a_1}{\partial t} + \frac{\gamma-1}{2}\frac{\partial u_1}{\partial x} = 0, \tag{3.53}$$

$$\frac{\partial u_1}{\partial t} + \frac{2}{\gamma-1}\frac{\partial a_1}{\partial x} = 0, \tag{3.54}$$

$$u_1(0,t) = f'(t), t \geqslant 0, \tag{3.55}$$

$$u_1(x,0) = 0, a_1(x,0) = 0,$$

$$\frac{\partial a_2}{\partial t} + \frac{\gamma-1}{2}\frac{\partial u_2}{\partial x} = -u_1\frac{\partial a_1}{\partial x} - \frac{\gamma-1}{2}a_1\frac{\partial u_1}{\partial x}, \tag{3.56}$$

$$\frac{\partial u_2}{\partial t} + \frac{2}{\gamma-1}\frac{\partial a_2}{\partial x} = -u_1\frac{\partial u_1}{\partial x} - \frac{2}{\gamma-1}a_1\frac{\partial a_1}{\partial x}, \tag{3.57}$$

$$u_2(0,t) = (-u_1)_x(0,t)f(t), t \geqslant 0,$$

$$u_2(x,0) = a_2(x,0) = 0. \tag{3.58}$$

a_1(or u_1) may be eliminated from eqs. (3.53)–(3.54) to obtain the wave equation

$$\frac{\partial^2 u_1}{\partial t^2} - \frac{\partial^2 u_1}{\partial x^2} = 0. \tag{3.59}$$

The (piston) boundary condition (3.55) dictates that only the forward wave solution of eq. (3.59) is relevant so that we have

$$u_1(x,t) = F(t-x) \tag{3.60}$$

$$= f'(t-x). \tag{3.61}$$

The solution of the initial boundary value problem, to first order, is

$$u_1(x,t) = \begin{cases} f'(t-x), & t > x, \quad (3.62) \\ 0, & t < x. \quad (3.63) \end{cases}$$

The integration of eq. (3.53) or eq. (3.54), after substitution of eq. (3.61), gives

$$a_1 = \begin{cases} \frac{\gamma-1}{2}f'(t-x), & t > x, \quad (3.64) \\ 0, & t < x. \quad (3.65) \end{cases}$$

Eqs. (3.62)–(3.65) confirm that, to this order, we have a simple wave with a constant value of the Riemann invariant along $t - x = \text{constant}$ (cf. eqs. (2.9) and (2.17)). Substituting the first order solution into the right sides of eqs. (3.56)–(3.57), we have

$$\frac{\partial a_2}{\partial t}+\frac{\gamma-1}{2}\frac{\partial u_2}{\partial x}=\frac{\gamma^2-1}{4}f'(t-x)f''(t-x), \quad t-x>0, \tag{3.66}$$

$$\frac{\partial u_2}{\partial t}+\frac{2}{\gamma-1}\frac{\partial a_2}{\partial x}=\frac{\gamma+1}{2}f'(t-x)f''(t-x), \quad t-x>0, \tag{3.67}$$

$$u_2(0,t)=f(t)f''(t), \quad t>0, \quad u_2(x,0)=a_2(x,0)=0. \tag{3.68}$$

The combination $a_2(x,t)-\frac{1}{2}(\gamma-1)u_2(x,t)$, as governed by eqs. (3.66)–(3.67), is easily verified to be constant along the positive characteristics $t-x=$ constant showing that the solution is a simple wave to even second order in ε. If we eliminate a_2 from eqs. (3.66)–(3.67), we can find a particular solution satisfying the inhomogeneous second order equation in u_2 and the boundary conditions in (3.68),

$$u_2(x,t)=\frac{\gamma+1}{2}xf'(\tau)f''(\tau)+f(\tau)f''(\tau), \quad \tau=t-x>0, \tag{3.69}$$

so that

$$a_2(x,t)=\frac{\gamma^2-1}{4}xf'(\tau)f''(\tau)+\frac{\gamma-1}{2}f(\tau)f''(\tau),$$
$$\tau=t-x>0, \tag{3.70}$$

using the constancy of the Riemann invariant along $\tau=$ constant or by direct substitution of $a_2-\frac{1}{2}(\gamma-1)u_2(x,t)=$ constant in eq. (3.66) and integration. We can now combine the first and second order solutions to write

$$u(x,t;\varepsilon)=\varepsilon f'(\tau)+\varepsilon^2\left\{f(\tau)f''(\tau)\right.$$
$$\left.+\frac{\gamma+1}{2}xf'(\tau)f''(\tau)\right\}+O(\varepsilon^3), \tag{3.71}$$

$$a(x,t;\varepsilon)=1+\frac{\gamma-1}{2}\varepsilon f'(\tau)+\frac{\gamma-1}{2}\varepsilon^2\left\{f(\tau)f''(\tau)\right.$$
$$\left.+\frac{\gamma+1}{2}xf'(\tau)f''(\tau)\right\}+O(\varepsilon^3). \tag{3.72}$$

We see from eqs. (3.71) and (3.72) that the terms of order ε^2 become comparable to those of order ε when

$$\varepsilon x=O(1) \tag{3.73}$$

which, near the front $t - x = 0$, separating the disturbed and the undisturbed states, is also equivalent to

$$\varepsilon t = O(1). \tag{3.74}$$

Thus, the non-uniformity of the regular perturbation scheme has been clearly demonstrated. Kevorkian and Cole (1981) proceed to find a uniformly valid solution by introducing the slow variable $\tilde{x} = \varepsilon x$ and arrive at the equation

$$\tilde{u}_{\tilde{x}} - \frac{\gamma + 1}{2}\tilde{u}\tilde{u}_{\tau} = 0. \tag{3.75}$$

This equation, after allowing for the change of sign in the second term due to the definition of τ etc., agrees with the inviscid form of the Burgers equation (2.44).

As we remarked earlier, we envisage a physical situation arising, say, out of a spherical explosion, and look at the phase of the wave when the initial energy has been largely spent, and the wave has evolved into an N shape (see Whitham (1952)). By this time, the wave has weakened considerably, yet it has small but finite amplitude, and suffers decay also due to geometrical spreading and diffusive damping. The latter effects are also small. Thus, we simulate these conditions as though they arise out of the small expansion of a piston, having a large radius $\hat{r}_p(\hat{t})$. The non-dimensional amplitude of the piston motion may be taken as ε so that

$$\frac{\mathrm{d}\hat{r}_{\mathrm{p}}}{\mathrm{d}\hat{t}} = \varepsilon a_* f\left(\frac{a_* \hat{t}}{l}\right)$$

or

$$\frac{\mathrm{d}}{\mathrm{d}t} r_{\mathrm{p}}(t) = \varepsilon f(t), \tag{3.76}$$

say, where l is the typical wavelength and $t = (a_* \hat{t}/l)$ is the non-dimensional time. A star denotes ambient conditions while '^' indicates dimensional quantities. The other small parameters of the problem are the inverse of the Reynolds number,

$$\mathrm{Re}^{-1} = \frac{\mu_*}{\rho_* a_* l} \equiv \mu_1, \tag{3.77}$$

and

$$x_{\mathrm{p}}^{-1} = \frac{l}{\hat{r}_{\mathrm{p}}(0)} = \mu_2, \tag{3.78}$$

representing diffusive and geometric effects, respectively.

These small effects introduce three physical time scales

$$\hat{t} = l \bigg/ \left(\frac{\mathrm{d}\hat{r}_{\mathrm{p}}}{\mathrm{d}\hat{t}} \right)_{\max}, \quad \hat{t}_2 = \frac{\rho_* l^2}{\mu_*}, \quad \hat{t}_3 = \frac{\hat{r}_{\mathrm{p}}(0)}{a_*}. \tag{3.79}$$

These give, respectively, (1) the time it takes the wave to propagate a distance l at the maximum perturbation velocity, (2) the time it takes the wave to diffuse a distance l, and (3) the time it takes the wave to travel a distance $\hat{r}_{\mathrm{p}}(0)$. The phenomena may be such that the time scales of these effects are different so that one has to consider them separately, or in different combinations (see Leibovich and Seebass (1974)). For our purpose we assume that all of them are of the same order and have to be accounted for concurrently. We closely follow Leibovich and Seebass (1974). We begin our analysis with the system (2.19)–(2.21), by ignoring the smallest terms shown in curly brackets, and including the terms due to spherical or cylindrical symmetry in the equation of continuity. Accordingly, we have the system

$$\hat{\rho}_{\hat{t}} + \hat{v}\hat{\rho}_{\hat{r}} + \hat{\rho}\hat{v}_{\hat{r}} = -\frac{j}{\hat{r}}\hat{\rho}\hat{v}, \tag{3.80}$$

$$\hat{v}_{\hat{t}} + \hat{v}\hat{v}_{\hat{r}} + \frac{1}{\hat{\rho}}\hat{p}_{\hat{r}} = \left(\frac{\frac{4}{3}\mu_0 + \mu_{v0}}{\rho_0} \right) \frac{\partial^2 \hat{v}}{\partial \hat{r}^2}, \tag{3.81}$$

$$\hat{s}_{\hat{t}} + \hat{v}\hat{s}_{\hat{r}} = \frac{k_0}{\hat{\rho}\hat{T}} \frac{\partial^2 T}{\partial \hat{r}^2}. \tag{3.82}$$

j in eq. (3.80) assumes values 2, 1 and 0 for spherical, cylindrical and plane symmetry, respectively. Eq. (3.82) replaces eq. (2.21) so that the dependent variables now are density, particle velocity and entropy. With the help of the gas law and the equation of state

$$p = R\rho T, \tag{3.83}$$

$$\left(\frac{p}{p_*} \right) = \left(\frac{\rho}{\rho_*} \right)^{\gamma} \mathrm{e}^{(s - s_*)/c_v}, \tag{3.84}$$

we may eliminate the pressure derivative term from eq. (3.81) and express the temperature derivative in eq. (3.82) in terms of entropy and density derivatives. We now introduce non-dimensional variables via the relations

$$\hat{\rho} = \rho_*(1 + \varepsilon\rho), \quad \hat{v} = \varepsilon a_* v, \quad \hat{s} = s_* + \varepsilon c_p \sigma \tag{3.85}$$

$$\hat{r} = xl, \quad \hat{t} = \frac{l}{a_*} t,$$

where ε is the amplitude of the wave. There is another small parameter $x_{\text{p}}^{-1} = l/r_{\text{p}}(0)$ which characterises the initial motion due to the piston with a large radius (see eq. (3.78)). The variables ρ, v and σ are of $O(1)$.

Eqs. (3.80)–(3.82) can be transformed through eqs. (3.83)–(3.85) into the following system in matrix form,

$$u_t + C(u)u_x = \text{Re}^{-1}D_1(u) + x_{\text{p}}^{-1}D_2(u, x). \tag{3.86}$$

where

$$\left.\begin{aligned} u &= \begin{pmatrix} \rho \\ v \\ \sigma \end{pmatrix}, \quad C(u) = \begin{pmatrix} \varepsilon v & 1+\varepsilon\rho & 0 \\ \dfrac{1+\varepsilon T}{1+\varepsilon\rho} & \varepsilon v & 1+\varepsilon T \\ 0 & 0 & \varepsilon v \end{pmatrix}, \\ D_1(u) &= \begin{pmatrix} 0 & 0 & 0 \\ 0 & \frac{4}{3}+\nu & 0 \\ \dfrac{\gamma-1}{\text{Pr}} & 0 & \dfrac{\gamma}{\text{Pr}} \end{pmatrix} u_{xx}, \\ D_2(u) &= -\frac{jx_{\text{p}}}{x}\begin{pmatrix} [1+\varepsilon\rho]v \\ 0 \\ 0 \end{pmatrix}. \end{aligned}\right\} \tag{3.87}$$

The non-dimensional temperature T is related to (non-dimensional) entropy and density by

$$\mathrm{e}^{\varepsilon\gamma\sigma} = \frac{1+\varepsilon T}{(1+\varepsilon\rho)^{\gamma-1}}. \tag{3.88}$$

In (3.87), $\text{Pr} = \mu c_p/k_0$ is the Prandtl number, $\mu = \mu_{v0}/\mu_0$ is the ratio of shear and dilational coefficients of viscosity and γ is the ratio of specific heats. All these coefficients are of order unity.

We wish to derive an equation describing waves on the constant state

$$u_{\text{c}} = \begin{pmatrix} 1 \\ 0 \\ 1 \end{pmatrix}. \tag{3.89}$$

Obviously, $u = u_{\text{c}}$ is a solution of the unperturbed form of eq. (3.86) with all small parameters equated to zero. We look for a perturbation solution of eq. (3.86) in the form

$$u = u_{\text{c}} + \varepsilon(u_0 + \varepsilon u_1 + \cdots + \mu_1 u_2 + \mu_2 u_3 + \cdots) \tag{3.90}$$

where μ_1 and μ_2 are Re^{-1} and x_{p}^{-1}, respectively, and constitute other small parameters, besides ε, arising from small viscosity and geometrical spreading. Substituting the expansion for the matrix

$$C(u_c + \varepsilon[u_0 + \cdots]) = C_0 + \varepsilon C_1(u_0, c) + \cdots, \qquad (3.91)$$
$$C_0 = C(u_c),$$

in eq. (3.86), we have, to different orders,

$$(u_0)_t + C_0(u_0)_x = 0, \qquad (3.92)$$
$$(u_1)_t + C_0(u_1)_x = -C_1(u_0, u_c)(u_0)_x, \qquad (3.93)$$
$$(u_2)_t + C_0(u_2)_x = D_1(u_c, u_0, x, t),$$
$$(u_3)_t + C_0(u_3)_x = D_2(u_c, u_0, x, t). \qquad (3.94)$$

Eqs. (3.92)–(3.94) follow from eq. (3.86) by equating terms of order $\varepsilon, \varepsilon^2$, $\varepsilon\mu_1$ and $\varepsilon\mu_2$, respectively. We may easily identify that

$$C_0 = \begin{pmatrix} 0 & 1 & 0 \\ 1 & 0 & 1 \\ 0 & 0 & 0 \end{pmatrix}, \quad C_1 = \begin{pmatrix} v & \rho & 0 \\ T-\rho & v & T \\ 0 & 0 & v \end{pmatrix}. \qquad (3.95)$$

Eq. (3.92) admits solutions of the type

$$u_0 = r_i U(x - \lambda_i t) \qquad (3.96)$$

if

$$[(C_0 - \lambda_i I)r_i]U_{X_i} = 0, \qquad (3.97)$$

where $X_i = x - \lambda_i t$, requiring r_i to be the right eigenvectors of C_0 corresponding to the eigenvalues $\lambda_i; U(X_i)$ is a scalar function of X_i. The matrix C_0 has eigenvalues $(+1, -1, 0)$ with corresponding right (transposed) eigenvectors $(1, 1, 0), (1, -1, 0)$, and $(-1, 0, 1)$, respectively. The eigenvalues give right running waves, left running waves and entropy waves, respectively. We assume that the boundary/initial conditions are such that only the right running waves are relevant (cf. earlier illustration of inviscid flow, eq. (2.17)). Thus, to the lowest order, $\lambda = 1, u_0 = rU(x-t) = rU(X)$ where $r^{\mathrm{T}} = (1, 1, 0)$. Substituting this solution into eq. (3.93) and changing to $X = x - t$ and t as independent variables, we have

$$(u_1)_t + (C_0 - I)(u_1)_X = -C_1(u_0, u_c)(u_0)_X. \qquad (3.98)$$

If we multiply eq. (3.98) with the left eigenvector l of the matrix C_0, that is, the vector l satisfying $l(C_0 - I) = 0$, we get

$$l(u_1)_t = -lC_1(u_0, u_c)(u_0)_X. \qquad (3.99)$$

The right hand side of (3.99) is a function of X only so that its integration leads to

$$lu_1 = -tlC_1(u_0)_X, \tag{3.100}$$

a secular term varying linearly with t and rendering the expansion (3.90) non-uniform when $t = O(1/\varepsilon)$. Similar non-uniformities will arise due to the expansions with respect to the parameters μ_1 and μ_2. The situation is retrieved by introducing dependence of the solution on the slow variables

$$\tau_1 = \varepsilon t, \quad \tau_2 = \mu_1 t, \quad \tau_3 = \mu_2 t, \tag{3.101}$$

besides the fast variable $t = \tau_0$, say. Thus, we have, formally,

$$\frac{\partial}{\partial t} = \frac{\partial}{\partial \tau_0} + \varepsilon \frac{\partial}{\partial \tau_1} + \mu_1 \frac{\partial}{\partial \tau_2} + \mu_2 \frac{\partial}{\partial \tau_3}. \tag{3.102}$$

Substituting (3.102) in eq. (3.86) and equating various order terms lead to

$$(u_1)_{\tau_0} + (C_0 - I)(u_1)_X = -C_1(u_0)_X - (u_0)_{\tau_1}, \tag{3.103}$$

$$(u_{k+1})_{\tau_0} + (C_0 - I)(u_{k+1})_X = D_k(u_c, u_0, X, \tau_0) - (u_0)_{\tau_{k+1}}, \quad k = 1, 2. \tag{3.104}$$

Multiplying these equations by the left eigenvector l will lead, on integration, to secular terms unless we put the right hand sides, so obtained, equal to zero. We thus have

$$lC_1 rU_X + lrU_{\tau_1} = 0, \tag{3.105}$$

$$lD_k - lrU_{\tau_{k+1}} = 0, \quad k = 1, 2, \tag{3.106}$$

where we have substituted for u_0 the solution, $u_0 = U(X, \tau_1, \tau_2, \tau_3)$; the function U is the solution of eq. (3.92) with $t = \tau_0$, and is now also a function of τ_1, τ_2 and τ_3. The dependence of U on τ_1, τ_2 and τ_3 is yet to be determined. This may be done by imposing the 'orthogonality' conditions (3.105)–(3.106) on this function. Explicit calculations give

$$\rho_0 = U(x - t, \tau_1, \tau_2, \tau_3), \quad u_0 = U(x - t, \tau_1, \tau_2, \tau_3), \tag{3.107}$$

$$\sigma_0 = 0,$$

so that, with eq. (3.88) in view, we have from (3.95)

$$C_1(u_0) = \begin{pmatrix} 1 & 1 & 0 \\ \gamma - 2 & 1 & \gamma - 1 \\ 0 & 0 & 1 \end{pmatrix} U(X, \tau_1, \tau_2, \tau_3). \tag{3.108}$$

The vector $l^{\mathrm{T}} = (1, 1, 1)$ satisfies $l(C_0 - I) = 0$. Therefore,

$$lr = 2,$$

$$lD_2 = -j\frac{x_{\mathrm{p}}}{x}(1,1,1)\begin{pmatrix} v \\ 0 \\ 0 \end{pmatrix} = -j(\tau_3 + x_{\mathrm{p}}^{-1}X)^{-1}U,$$

$$lC_1 r = (\gamma+1)U, \quad lD_1 = \left(\tfrac{4}{3} + \nu + \frac{\gamma-1}{\mathrm{Pr}}\right)U_{XX}. \tag{3.109}$$

Substituting (3.107)–(3.109) etc. into eqs. (3.105)–(3.106), we have

$$U_{\tau_1} + \Gamma U U_X = 0, \tag{3.110}$$

$$U_{\tau_2} - \frac{\kappa}{2}U_{XX} = 0, \tag{3.111}$$

$$U_{\tau_3} + \frac{j}{2}\frac{U}{\tau_3} = 0, \tag{3.112}$$

where

$$\Gamma = \frac{\gamma+1}{2}, \quad \kappa = \tfrac{4}{3} + \nu + \frac{\gamma-1}{\mathrm{Pr}}, \tag{3.113}$$

and $x_{\mathrm{p}}^{-1}X$ has been neglected in comparison with τ_3 in writing lD_2. If we multiply eqs. (3.110)–(3.112) by ε, μ_1 and μ_2, respectively, and add, we get, using (3.102) and (3.107), the reconstituted equation

$$U_t + \varepsilon\Gamma U U_X + \frac{j}{2t}U = \frac{\kappa}{2}\mathrm{Re}^{-1}U_{XX}. \tag{3.114}$$

This is the generalised Burgers equation which includes the effects of spherical or cylindrical spreading, besides those of nonlinearity and diffusion. Eq. (3.110) is the same as derived by Kevorkian and Cole (1981) (see eq. (3.75)). Eqs. (3.110)–(3.112) may be valid individually or in certain combinations in different time regimes. For a discussion of this aspect and for the corresponding solutions of these equations, reference may be made to Leibovich and Seebass (1974). If we write $\tau = \varepsilon\Gamma t$ and $\delta = \kappa/\varepsilon\,\mathrm{Re}\,\Gamma$, eq. (3.114) can be transformed into

$$U_\tau + U U_X + \frac{jU}{2\tau} = \frac{\delta}{2}U_{XX}. \tag{3.115}$$

Henceforth, we again write t in place of τ in the statement of eq. (3.115).

3.4 Singular perturbation solution for non-planar N waves

We now discuss the initial value problem for eq. (3.115) with

$$U(X, t_0) = \begin{cases} U_0 X/l_0, & |X| < l_0, \\ 0, & |X| > l_0, \end{cases} \tag{3.116}$$

so that, at $t = t_0$, we have a sawtooth structure of an N wave with an amplitude equal to U_0 and wavelength l_0. We give here a first order matched asymptotic solution of this problem due to Crighton and Scott (1979). The procedure is similar to that in sec. 3.2. There are some advantages in transforming eqs. (3.115)–(3.116) according to

$$V = \left(\frac{t}{t_0}\right)^{j/2} \frac{U}{U_0}, \quad x = \frac{X}{l_0},$$

$$T = \begin{cases} 1 + U_0(t - t_0)/l_0, & j = 0, \\ 1 + 2U_0(t_0^{1/2} t^{1/2} - t_0)/l_0, & j = 1, \\ 1 + (U_0 t_0/l_0) \ln(t/t_0), & j = 2, \end{cases} \tag{3.117}$$

so that initial value problem (3.115)–(3.116) now becomes

$$\frac{\partial V}{\partial T} + V \frac{\partial V}{\partial x} = \varepsilon g(T) \frac{\partial^2 V}{\partial x^2}, \tag{3.118}$$

$$V(x, 1) = \begin{cases} x, & |x| < 1, \\ 0, & |x| > 1, \end{cases} \tag{3.119}$$

where ε is an inverse Reynolds number given by

$$\varepsilon = \begin{cases} \delta/2U_0 l_0, & j = 0, \\ \delta/U_0 l_0 T_0, & j = 1, \\ \delta\, \mathrm{e}^{-1/T_0}/2U_0 l_0, & j = 2, \end{cases} \tag{3.120}$$

with

$$T_0 = \begin{cases} 2U_0 t_0/l_0, & j = 0, \\ 2U_0 t_0/j l_0, & j = 1, 2. \end{cases} \tag{3.121}$$

The function $g(T)$ is given by

$$g(T) = 1, \quad \tfrac{1}{2}(T + T_0 - 1), \quad \mathrm{e}^{T/T_0} \tag{3.122}$$

for $j = 0, 1, 2$, respectively. In view of the absence of the factor $(\gamma + 1)/2$ in eq. (3.115), our scalings are slightly different from those of Crighton and Scott (1979).

The form (3.118), with variable viscosity, makes the outer solution

$$V(x,T)=\begin{cases}\dfrac{x}{T}, & x<T^{1/2},\\[2ex] 0, & x>T^{1/2},\end{cases} \tag{3.123}$$

the same for all $j=0,1,2$. In view of eq. (3.117), the solution is required for much smaller T than t for $j=1,2$. This is a significant gain if the solution is to be found numerically for very large time.

A formal outer expansion for eq. (3.118) immediately shows that all higher order terms $O(\varepsilon^n)$, subject to zero initial conditions in eq. (3.119), are zero. The shock discontinuity according to the zero order outer solution has the locus $x=T^{1/2}$, which is easily obtained by writing the conservative form

$$\frac{\partial}{\partial T}(V)+\frac{\partial}{\partial x}(\tfrac{1}{2}V^2)=0, \tag{3.124}$$

so that

$$\frac{\mathrm{d}x}{\mathrm{d}T}=\frac{[\frac{1}{2}V^2]}{[V]}=\frac{1}{2}\frac{x}{T}, \tag{3.125}$$

leading by integration to $x=T^{1/2}$. This satisfies the initial condition $x=1$ at $T=1$. The notation '[ ]' denotes the jump of the bracketed quantity across the shock.

To obtain the inner solution, we introduce the stretched variable

$$x^*=\frac{x-T^{1/2}}{\varepsilon}, \tag{3.126}$$

with the inviscid shock as the origin, so that eq. (3.118) becomes

$$\varepsilon\frac{\partial V}{\partial T}-\frac{1}{2}\frac{1}{T^{1/2}}\frac{\partial V}{\partial x^*}+V\frac{\partial V}{\partial x^*}=g(T)\frac{\partial^2 V}{\partial x^{*2}}. \tag{3.127}$$

The solution is sought in the form

$$V=V_0^*(x^*,T)+\varepsilon V_1^*(x^*,T)+\cdots. \tag{3.128}$$

Substitution of (3.128) in eq. (3.127) shows that $V_0^*(x^*,T)$ and $V_1^*(x^*,T)$ are governed by

$$\left(V_0^*-\frac{1}{2T^{1/2}}\right)\frac{\partial V_0^*}{\partial x^*}=g(T)\frac{\partial^2 V_0}{\partial x^{*2}}, \tag{3.129}$$

$$g(T)\frac{\partial^2 V_1^*}{\partial x^{*2}} + (\tfrac{1}{2}T^{-1/2} - V_0^*)\frac{\partial V_1^*}{\partial x^*} - V_1^*\frac{\partial V_0^*}{\partial x^*} = \frac{\partial V_0^*}{\partial T}. \tag{3.130}$$

It is easy to verify that eq. (3.129) has the solution

$$V_0^* = \tfrac{1}{2}T^{-1/2}\left[1 - \tanh\frac{x^* - A(T)}{4T^{1/2}g(T)}\right], \tag{3.131}$$

matching, to zero order, the outer solution (3.123) expressed in terms of the inner variable, $V_0 = 1/T^{1/2} + \varepsilon x^*/T$, as $\varepsilon \to 0$ and $x^* \to -\infty$, and tending to zero as $x^* \to +\infty$, according to intermediate limits. The function $A(T)$, which arises as a 'constant' of integration from eq. (3.129), has to be determined from higher order approximations, and would give the shock displacement due to viscous diffusion. If we change the independent variables in eq. (3.130) to

$$y = \frac{x^* - A(T)}{4T^{1/2}g(T)} \tag{3.132}$$

and T, we get

$$\frac{1}{2}\frac{\partial^2 V_1^*}{\partial y^2} + \tanh y\frac{\partial V_1^*}{\partial y} + \operatorname{sech}^2 y\, V_1^* = 8Tg(T)\frac{\partial V_0^*}{\partial T}. \tag{3.133}$$

Two linearly independent solutions of the homogeneous part of the 'ordinary' differential equation (3.133) are $\operatorname{sech}^2 y$ and $y\operatorname{sech}^2 y + \tanh y$, so that a particular solution of eq. (3.133) may be found to be

$$\begin{aligned} V_1^* = {} & \frac{\mathrm{d}A}{\mathrm{d}t} - T^{-1/2}g(T)(y^2\operatorname{sech}^2 y + 2y\tanh y - 1 - 2y) \\ & + G(T)(y\operatorname{sech}^2 y + \tanh y) + K(T)\operatorname{sech}^2 y \\ & + 4T^{1/2}\frac{\mathrm{d}g}{\mathrm{d}T}\{y - (\ln\cosh y)\tanh y + \tanh y \\ & + \operatorname{sech}^2 y[y(\ln 2 + \tfrac{1}{2}y) + \tfrac{1}{2}\operatorname{diln}(1 + e^{2y})]\} \end{aligned} \tag{3.134}$$

where

$$\operatorname{diln}(x) = -\int_1^x \frac{\ln t}{t-1}\mathrm{d}t$$

(Abramowitz & Stegun, 1964, p. 1004). The functions $g(T)$ and $K(T)$ are to be determined by matching. The solution V_1^* is to be matched to $V_1 = 0$ in the intermediate limit, that is, as $\varepsilon \to 0$, $y \to -\infty$ when $x < T^{1/2}$, and to $V_1 = 0$ as $\varepsilon \to 0$, $y \to +\infty$, when $x > T^{1/2}$. Hence we get the following limiting equations from eq. (3.134):

$$\frac{\mathrm{d}A}{\mathrm{d}t} - AT^{-1} + T^{-1/2}g(T) - G(T) - 4T^{1/2}\frac{\mathrm{d}g}{\mathrm{d}T}(1 + \ln 2) = 0, \tag{3.135}$$

$$\frac{\mathrm{d}A}{\mathrm{d}t} + T^{-1/2}g(T) + G(T) + 4T^{1/2}\frac{\mathrm{d}g}{\mathrm{d}T}(1 + \ln 2) = 0. \tag{3.136}$$

Elimination of $G(T)$ leads to

$$\frac{\mathrm{d}A}{\mathrm{d}T} + T^{-1/2}g(T) - \tfrac{1}{2}AT^{-1} = 0 \tag{3.137}$$

which, on integration, gives

$$A(T) = A(1) - T^{1/2}\int_1^T \frac{g(t)}{t}\,\mathrm{d}t. \tag{3.138}$$

Crighton and Scott have matched the above solution to the embryonic shock region over the time the discontinuous shock adjusts to a steady Taylor shock and have, thereby, found $A(1) = 0$. The substitution of $g(T)$ from eq. (3.122) into eq. (3.138) yields the explicit form of $A(T)$:

$$A(T) = -T^{1/2}\ln T \quad \text{(planar)}, \tag{3.139}$$

$$A(T) = -\tfrac{1}{2}T^{1/2}\{T - 1 + (T_0 - 1)\ln T\} \quad \text{(cylindrical)}, \tag{3.140}$$

$$A(T) = -T^{1/2}\left\{\mathrm{Ei}\left(\frac{T}{T_0}\right) - \mathrm{Ei}(T_0^{-1})\right\} \quad \text{(spherical)}, \tag{3.141}$$

where Ei is the exponential integral defined by

$$\mathrm{Ei}(x) = \int_{-\infty}^{x} t^{-1}\,\mathrm{e}^t\,\mathrm{d}t$$

(see Abramowitz and Stegun (1964, p. 228)).

Crighton and Scott show that, for planar and cylindrical cases, nothing short of the knowledge of the full solution would be required for further progress, while for the spherical case they are able to find the constant in the final linear solution of the (entirely) diffusive regime. We note, *en passant*, that the N wave solution of the linearised form of eq. (3.115) can easily be found to be

$$U = \left(\frac{t_\mathrm{f}}{t}\right)^{(j+1)/2}\frac{A_\mathrm{f}}{2\delta}\frac{X}{t}\,\mathrm{e}^{-X^2/2\delta t}$$

$$\equiv CXt^{-(j+3)/2}\,\mathrm{e}^{-X^2/2\delta t}, \quad \text{say}, \tag{3.142}$$

where A_f is the area $\int_{-\infty}^{\infty} U(x, t_f)\mathrm{d}x$ and t_f is the time when the profile has propagated far enough to become several times its original length. The constant C will vary for different geometries. For the spherical case, it was found by Crighton and Scott (1979) by the matching of another perturbation solution in the shock region to the lossless outer solution; for the cylindrical case, it remains to be determined (see sec. 5.6 for evaluation of C from the numerical solution).

The matched asymptotic solution found above becomes non-uniform for large T due to the failure of one of the following hypotheses underlying the perturbation analysis, namely (i) the shock displacement due to diffusion remains small, (ii) the shock remains thin in comparison with the characteristic scale of the inviscid flow, (iii) the Taylor-type shock solution remains valid in the shock region. Before we summarise the conclusions of Crighton and Scott regarding the validity of the perturbation solution for different geometries, we note that the embryo shock region which immediately follows the initial establishment of a (discontinuous) N wave was found by Crighton and Scott for $j = 0, 1, 2$; for the planar case, the exact solution is known, but it does not satisfy the discontinuous initial condition (3.116).

For the planar case, the shock, according to weak (inviscid) shock theory, is located at $x = T^{1/2}$. According to eq. (3.139), the diffusive effects spread the shock away from its original 'weak theory' position by the time $\ln T = O(\varepsilon^{-1})$. Further, the ratio $(\varepsilon V_1^*/V_0^*)$ in the limit $T \to \infty$, with y fixed, shows that the Taylor solution will not be valid when $\ln T = O(\varepsilon^{-1})$ (see eqs. (3.131)–(3.134)). Thus, the perturbation solution breaks down, the steady Taylor solution implicit in eq. (3.131) is not valid and the shock is no longer thin. Fortunately, from the Hopf–Cole transformation, we know the complete solution (2.100) of the Burgers equation, that describes the evolution of the N wave all the way from a smooth Taylor shock to the old-age stage when the linear (heat) equation describes the flow and the wave has died down to a small amplitude.

For the cylindrical case, eq. (3.140), in the limit $T \to \infty$, shows that all the conditions (i)–(iii) are violated at the same time, $T = O(\varepsilon^{-1})$. This non-uniformity affects the whole wave and the complete equation (3.115) will have to be solved to make any headway. In the old-age regime, however, the linearised form of eq. (3.115) gives the dipole solution

$$U = C\frac{X}{t^2}\mathrm{e}^{-X^2/2\delta t}, \tag{3.142a}$$

for which the constant C cannot be found except via the numerical solution when the latter matches eq. (3.142a) (see sec. 5.6 for further details).

The situation is a little better for the spherical case $j = 2$. Taylor's shock solution with $A(T)$ given by eq. (3.141) continues to hold for some time. However, the combined effect of nonlinear convection and geometrical spreading in the flow is so intense that the balance between nonlinearity and diffusion in the Taylor shock is not sustained. Thus, the first non-uniformity in the perturbation solution takes place due to the failure of condition (iii). Crighton and Scott, while unable to treat this local non-uniformity, find a matched asymptotic solution in a subsequent time régime when the nonlinearity becomes recessive. Now, the linearised equation is solved uniquely in terms of the error function and matched to the inviscid N wave solution (3.123) near the node. This solution describes the large T behaviour of the evolutionary shock which is now of non-Taylor type and in which spherical spreading is a controlling mechanism. At this stage, condition (ii) is violated. The non-uniformity of the solution now affects the whole wave. A new solution of the linearised equation with rather unusual scalings is found, matching the inviscid N wave solution as well as the erf solution found earlier. This analysis leads to the explicit determination of the constant C in eq. (3.142) for $j = 2$. The point to emphasise is that the inviscid solution near the node remains valid for a very long time indeed, explaining the excellent agreement of the analytic expression for the Reynolds number, obtained on the basis of this assumption, with the numerical results. This was demonstrated by Sachdev and Seebass (1973) (see, however, sec. 5.6 for more recent work).

We shall refer again to this work in secs. 3.7 and 5.6 when we make a comparative study of the results discussed herein with those obtained via another analytic method and by numerical methods.

3.5 The periodic plane piston problem

Now we consider an initial boundary value problem arising from a sinusoidally oscillating piston. For this purpose we take another variant of the Burgers equation, due originally to Mendousse (1953), which was subsequently studied by several authors (Blackstock, 1964, Lesser & Crighton, 1975, Rudenko & Soluyan, 1977 and Parker, 1980). The equation has the normalised form

$$u_x - uu_\tau = \frac{\delta}{2} u_{\tau\tau}, \tag{3.143}$$

where x is the spatial variable as measured from the driving piston; τ is the retarded time equal to $t - x/a_0$, denoting the time elapsed since the passage of a reference wavelet. Eq. (3.143) can be derived from eq. (2.44) or directly by using perturbation techniques (see Lesser and Crighton (1975)). We emphasise that, in both this section and the next, all the details of the wave such as the Taylor-shock structure or 'old-age' behaviour will be referred to with respect to time τ for a given value of the distance x measured from the piston.

The normalised boundary condition for eq. (3.143) is

$$u(0, \tau) = \begin{Bmatrix} 0, & \tau < 0, \\ \sin \tau, & \tau \geqslant 0. \end{Bmatrix} \tag{3.144}$$

We ignore the effect of the finite displacement of the piston and consider the evolution of the wave produced by the signals sent out from $x = 0$ into the domain $x > 0$. Before solving eq. (3.143) subject to eq. (3.144), we consider the inviscid problem with $\delta = 0$ in eq. (3.143). The characteristics of this equation

$$\frac{\mathrm{d}x}{1} = \frac{\mathrm{d}\tau}{-u} = \frac{\mathrm{d}u}{0} \tag{3.145}$$

give the general solution

$$u = f(\sigma), \quad \sigma = \tau + xf(\sigma), \tag{3.146}$$

where σ is the characteristic variable, which may be interpreted as a distorted version of time τ, the distortion increasing with x. With the boundary condition (3.144), $f(\sigma) = \sin \sigma$ and the solution becomes

$$u = \sin \sigma, \quad \sigma = \tau + x \sin \sigma$$

or

$$u = \sin (\tau + xu). \tag{3.147}$$

This system describes the solution parametrically until $x = 1$, when the shock is formed. To see this, we differentiate the first form of eq. (3.147) with respect to σ to obtain

$$\frac{\mathrm{d}\tau}{\mathrm{d}\sigma} = 1 - x \cos \sigma. \tag{3.148}$$

This expression becomes zero first when $x = 1/(\cos \sigma)_{\max} = 1$. So the solution (3.147) remains valid until $x = 1$, when it develops a shock. From this stage, the evolution of this solution with distance should, in fact, be studied with full Burgers equation until the shock structure (with respect to τ) becomes Taylor-type. The solution (3.147) can be put into a more familiar form, the so-called Fubini's solution (Fubini, 1935), if we write

$$u = \sum_{n=1}^{\infty} B_n(x) \sin n\tau, \tag{3.149}$$

where

$$B_n = \frac{2}{\pi} \int_0^{\pi} u \sin n\tau \, \mathrm{d}\tau \tag{3.150}$$

$$= \frac{2}{\pi} \int_0^{\pi} \sin \xi \sin (n\xi - nx \sin \xi)(1 - x \cos \xi) \mathrm{d}\xi, \tag{3.151}$$

by putting $\xi = \tau + xu$ and using eq. (3.147). Changing in eq. (3.151) the products of trigonometric functions into sums of cosine functions and using the formula

$$\frac{1}{\pi} \int_0^{\pi} \cos (\nu\xi - nx \sin \xi) \, \mathrm{d}\xi = J_\nu(nx), \tag{3.152}$$

together with recurrence relations for Bessel functions, we get

$$B_n = 2\frac{J_n(nx)}{nx}. \tag{3.153}$$

The Fourier series form of u, eq. (3.149), is the Fubini solution

$$u = \sum_{n=1}^{\infty} 2\frac{J_n(nx)}{nx} \sin n\tau \tag{3.154}$$

so that we have a means of knowing how higher harmonics are generated by the quadratic nonlinearity in eq. (3.143) with $\delta = 0$.

We now pursue the boundary value problem (3.143)–(3.144) (which is, in fact, an initial value problem for the parabolic equation (3.143)), and continue the study of sec. 2.8, with minor differences due to change of sign in the nonlinear term $-uu_\tau$ (Blackstock, 1964). The Hopf–Cole transformation

$$u = \delta[\log \zeta]_\tau, \tag{3.155}$$

that is,

$$\zeta = \exp\left[\delta^{-1} \int_{-\infty}^{\tau} u \, \mathrm{d}\tau \right], \tag{3.156}$$

reduces eq. (3.143) to

$$\zeta_x = \frac{\delta}{2}\zeta_{\tau\tau} \tag{3.157}$$

with initial condition

$$\zeta_0 = \zeta(0,\tau) = \begin{cases} e^{\delta^{-1}(1-\cos\tau)}, & \tau > 0, \\ 1, & \tau < 0. \end{cases} \tag{3.158}$$

The solution of eqs. (3.157)–(3.158) is

$$\begin{aligned}
\zeta &= (2\pi\delta x)^{-1/2} \int_{-\infty}^{\infty} \zeta_0(\lambda) e^{-(\lambda-\tau)^2/2\delta x} \, d\lambda \\
&= (2\pi\delta x)^{-1/2} \left[-\int_0^{-\infty} \zeta_0(\lambda) e^{-(\lambda-\tau)^2/2\delta x} \, d\lambda \right. \\
&\quad \left. + \int_0^{+\infty} \zeta_0(\lambda) e^{-(\lambda-\tau)^2/2\delta x} \, d\lambda \right] \\
&= \tfrac{1}{2} \operatorname{erfc}(\tau/m) - \pi^{-1/2} e^{1/\delta} \int_{\tau/m}^{\infty} \exp\left[-q^2 - \frac{1}{\delta}\cos(mq-\tau) \right] dq \\
&\quad + \pi^{-1/2} e^{1/\delta} \int_{-\infty}^{\infty} \exp\left[-q^2 - \frac{1}{\delta}\cos(mq+\tau) \right] dq \\
&= I_1 + I_2 + I_3, \text{ say,}
\end{aligned} \tag{3.159}$$

where $m = (2\delta x)^{1/2}$. The integrals I_1 and I_2 approach zero as τ increases and represent transients. The main wave is given by I_3. If we employ the formula

$$e^{-a\cos\theta} = \sum_{n=0}^{\infty} \varepsilon_n (-1)^n I_n(a) \cos n\theta \tag{3.160}$$

in I_3, where ε_n are Neumann numbers, $\varepsilon_0 = 1, \varepsilon_n = 2$ for $n \geqslant 1$, and the standard integral $\int_0^\infty \cos nmq \, e^{-q^2} dq = \frac{1}{2}\sqrt{\pi} e^{-n^2m^2/4}$, we arrive at the solution (3.155) with

$$\zeta = \sum_{n=0}^{\infty} \varepsilon_n (-1)^n I_n\left(\frac{1}{\delta}\right) e^{-\delta n^2 x/2} \cos n\tau. \tag{3.161}$$

I_n in eqs. (3.160) and (3.161) stands for the Bessel function of order n with imaginary argument. Now, several interesting results follow. The asymptotic expression for Bessel function I_n for small δ is

$$I_n\left(\frac{1}{\delta}\right) \sim \frac{e^{1/\delta}}{(2\pi/\delta)^{1/2}} \left[1 - \frac{4n^2 - 1^2}{1!(8/\delta)} + \frac{(4n^2-1^2)(4n^2-3^2)}{2!(8/\delta)^2} + \cdots \right]. \tag{3.162}$$

In the following discussion, we shall ignore the factor $e^{1/\delta}/(2\pi/\delta)^{1/2}$ which will disappear through the transformation (3.155). If we substitute, for all I_n, the first term (unity) in the square brackets of (3.162) into eq. (3.161), then we recognize that the right hand side is a theta function. Therefore,

$$\zeta \approx \theta_4\left(\frac{\tau}{2}, \quad \mathrm{e}^{-x\delta/2}\right) \tag{3.163}$$

(see Abramowitz and Stegun 1964). This function has the property that

$$\left[\ln \theta_4\left(\frac{y}{2}, \mathrm{e}^{-\Omega}\right)\right]_y = \sum_{n=1}^{\infty} (\sinh n\Omega)^{-1} \sin ny. \tag{3.164}$$

Comparing (3.163) and (3.164) with eq. (3.155), we have

$$u \approx \delta \sum_{n=1}^{\infty} \frac{\sin n\tau}{\sinh \dfrac{nx\delta}{2}}. \tag{3.165}$$

The approximation has some validity if δ is very small; when n becomes of order $1/\delta$, the exponential factor in eq. (3.161) helps in keeping down the error. The expression (3.165) for u is however an exact solution of the Burgers equation (see sec. 2.9). A further exact solution may be obtained by rewriting the series in (3.162) as

$$I_n\left(\frac{1}{\delta}\right) \sim \mathrm{e}^{-n^2\delta/2}\left[1 + \frac{\delta}{8} - (n^2 - 9/32)\frac{\delta^2}{4} + \cdots\right]. \tag{3.166}$$

This rearrangement of a divergent series requires some justification (see Blackstock (1964)). However, a formal multiplication of the two factors on the right hand side of eq. (3.166) will confirm this step. Again, taking only the first term in (3.166) and substituting into eq. (3.161) lead to

$$\zeta \approx \theta_4\left(\frac{\tau}{2}, \quad \mathrm{e}^{-\delta(1+x)/2}\right) \tag{3.167}$$

so that we have, using (3.164) and (3.155),

$$u \approx \delta \sum_{n=1}^{\infty} \frac{\sin n\tau}{\sinh \dfrac{n\delta}{2}(1+x)}. \tag{3.168}$$

This is often referred to as Fay's solution (Fay, 1931) and is an exact solution of eq. (3.143). We further transform Fay's solution by noting that

$$(\sinh nf^{-1})^{-1} = f \sum_{k=0}^{\infty} \varepsilon_k \frac{(-1)^k n}{n^2 + k^2\pi^2 f^2}, \tag{3.169}$$

$$\pi - y = \sum_{n=1}^{\infty} \frac{2}{n} \sin ny, \quad 0 \leqslant y < 2\pi, \tag{3.170}$$

$$\frac{\pi \sinh a(\pi - y)}{\sinh a\pi} = \sum_{n=1}^{\infty} \frac{2n}{n^2 + a^2} \sin ny, \quad 0 \leqslant y < 2\pi, \tag{3.171}$$

(ε_k are again Neumann numbers).

Substitution of (3.169) into eq. (3.168) shows that the term corresponding to $k = 0$ leads to eq. (3.170), while the remaining terms lead to a sum involving the left hand side of eq. (3.171). Therefore, we get

$$u = \frac{1}{1+x}\left[\pi - \tau + 2\pi \sum_{k=1}^{\infty} (-1)^k \frac{\sinh[2k(\pi - \tau)/\Delta]}{\sinh(2\pi k/\Delta)}\right], \tag{3.172}$$

where $\Delta = \delta(1 + x)/\pi$. Since the arguments of the hyperbolic sines are large for small δ, we can approximate the series in eq. (3.172) by

$$\sum_{k=1}^{\infty} (-1)^k [\mathrm{e}^{-2k\tau/\Delta} - \mathrm{e}^{-2k(2\pi-\tau)/\Delta}]$$

$$= \frac{1}{2}\left[\tanh(\tau/\Delta) - \tanh\frac{2\pi - \tau}{\Delta}\right]. \tag{3.173}$$

Eq. (3.172) now becomes

$$u = \frac{1}{1+x}\left[\pi - \tau + \pi \tanh(\tau/\Delta) - \pi \tanh\frac{2\pi - \tau}{\Delta}\right]. \tag{3.174}$$

For τ away from 2π and small δ, the second hyperbolic tangent may be replaced by unity, and eq. (3.174) becomes

$$u = \frac{1}{1+x}\left[-\tau + \pi \tanh\frac{\pi\tau}{\delta(1+x)}\right]. \tag{3.174a}$$

On the other hand, for $\tau \sim 2\pi$, the first hyperbolic tangent is close to unity so that we may obtain eq. (3.174a) with τ replaced by $\tau - 2\pi$. Eq. (3.174a) represents another exact solution of Burgers' equation (cf. eq. (2.118)). Finally, in the limit $\delta \to 0$, we may replace (3.174a) by

$$u = \left\{\begin{array}{ll} \dfrac{1}{1+x}[-\tau + \pi], & 0 < \tau < \pi, \\ \dfrac{1}{1+x}[-\tau - \pi], & -\pi < \tau < 0, \end{array}\right\} \tag{3.175}$$

which is the (exact) sawtooth solution of the inviscid Burgers equation. Thus, it is remarkable that, despite several approximations in the derivations, we get several exact solutions of the Burgers equation, holding, of course, in different domains. This idiosyncratic behaviour of the Burgers equations will reappear when we consider the singular perturbation solution of this problem in the next section. In conclusion, we note that the solution (3.175) emerges as a non-viscous limit of the viscous solution and satisfies, correspondingly, the equation which is a non-viscous limit of the viscous equation – a rather uncommon situation.

Blackstock (1964) has improved upon Fay's solution by retaining more terms in the series in (3.162).

3.6 Singular perturbation analysis of the periodic piston problem

We now turn to the singular perturbation analysis of the periodic boundary value problem (3.143)–(3.144), after the shock is formed at $x = 1$. At this distance from the piston, the lossless solution becomes multi-valued for each value of $\tau = \sigma = 2k\pi$, where k is an integer, according to eq. (3.147). At each of these temporal points a shock forms and the solution (3.147) breaks down. However, this solution of eq. (3.143) preserves the asymmetry and periodicity of the boundary condition (3.144); any sharp shock transition remains at $\tau = 2k\pi$ for all values of $x > 1$. The evolution of the profile between $x = 1$, the point of inception of the shock, and the establishment of the quasi-steady Taylor shock with respect to τ (that is, the embryonic shock régime) can be traced only by solving the full Burgers eq. (3.143), since, in this régime, all terms are equally important. After the quasi-steady shock has been established, it is possible to solve the problem by singular perturbation methods. The thickness of the quasi-steady shock is $O(\delta)$, as follows easily from the balance of the terms uu_τ and $\frac{1}{2}\delta u_{xx}$ (Parker, 1980).

For $x = O(\mu^{-1})$, where $\delta \ll \mu \ll 1$, the inviscid solution (3.147), as in the earlier analysis of sec. 3.4, remains valid as the first term of an outer expansion, away from each of shock locations $\tau = 2k\pi$, and close to the 'nodes', $\tau = (2k+1)\pi$. It may be simplified by introducing the variables $\bar{x}$ and β,

$$x = \mu^{-1}\bar{x}, \quad \sigma - (2k+1)\pi = \mu\beta, \tag{3.176}$$

so that eq. (3.147) yields the outer solution to order μ:

$$\begin{aligned} u_0 &= \sin((2k+1)\pi + \mu\beta) = -\sin\mu\beta \sim -\mu\beta, \\ \tau &- (2k+1)\pi - \mu\beta = x\sin\mu\beta \sim x\mu\beta. \end{aligned} \tag{3.177}$$

Therefore, the outer solution, for $x = O(\mu^{-1})$, becomes

$$u_0 \sim -\mu\beta \sim -\frac{\tau-(2k+1)\pi}{x+1}, \tag{3.178}$$

holding in each interval $2k\pi < \tau < 2(k+1)\pi$. This is a straight ramp or sawtooth wave with discontinuities of strength $2\pi/(x+1)$ at each $\tau = 2k\pi$. We match this outer solution with an inner solution about the sharp shock discontinuities as in sec. 3.4. We define the inner stretched variables

$$\bar{\tau} = \mu\frac{(\tau - 2k\pi)}{(\delta/2)}, \quad \bar{u}(\bar{\tau}, \bar{x}) = \mu^{-1}u. \tag{3.179}$$

Eq. (3.143) becomes

$$-\bar{u}\bar{u}_{\bar{\tau}} = \bar{u}_{\bar{\tau}\bar{\tau}} - \mu^{-1}\frac{\delta}{2}\bar{u}_{\bar{x}}. \tag{3.180}$$

In view of $\delta \ll \mu \ll 1$, the lowest order equation for $\bar{u}$ becomes

$$-vv_\tau = v_{\tau\tau}. \tag{3.181}$$

This is to be matched to the lowest order outer solution (3.178), say, for $k = 0$. In this case, the shock is at $\tau = 0$ and the outer solution is $u \sim (\pi - \tau)/(x+1)$ for $0 < \tau < 2\pi$ and $u \sim -(\tau + \pi)/(x+1)$ for $-2\pi < \tau < 0$. Hence the solution to eq. (3.181) should satisfy the intermediate limits

$$v \to \mp\pi/(\bar{x} + \mu) \quad \text{as } \bar{\tau} \to \mp\infty. \tag{3.182}$$

Such a solution of eq. (3.181) is easily found to be

$$\bar{u} \sim v(\bar{\tau}, \bar{x}) = \frac{\pi}{\bar{x}+\mu}\tanh\frac{\pi\bar{\tau}}{2(\bar{x}+\mu)}. \tag{3.183}$$

This is a Taylor shock with strength $2\pi\mu/(\bar{x}+\mu)$. If we use the matching principle for getting a composite solution, namely

$$y_c = y_{out} + y_{in} - y_{match} \tag{3.184}$$

(see Bender and Orszag (1978)), we have

$$\begin{aligned} u \sim u_c(\tau, x, \delta) &= \frac{\pi-\tau}{x+1} + \frac{\mu\pi}{\bar{x}+\mu}\tanh\frac{\pi\mu\tau}{\delta(\bar{x}+\mu)} - \frac{\pi}{x+1} \\ &= \frac{1}{x+1}\left[\pi\tanh\frac{\pi\tau}{\delta(x+1)} - \tau\right]. \end{aligned} \tag{3.185}$$

This representation of the solution holds throughout the basic period $-\pi \leqslant \tau \leqslant \pi$, provided the assumption underlying the analysis, $\delta^{-1} \gg x \gg 1$, holds.

It is interesting to note that the inviscid Fubini solution (3.154) holding for $x \leqslant 1$ has now transformed into $(\pi - \tau)/(1 + x)$; of course, it holds away from the shock layer $\tau \sim 0$.

Lesser and Crighton (1975) have shown that the Fourier series form of the composite solution (3.185) produces Fay's solution. We proved the converse of this result earlier in sec. 3.5. Explicitly, if we write

$$u_c = \frac{1}{x+1}\left[\pi \tanh \frac{\pi\tau}{\delta(x+1)} - \tau\right] = \sum_{n=1}^{\infty} A_n(x,\delta) \sin n\tau, \quad |\tau| < \pi, \tag{3.186}$$

then

$$A_n(x,\delta) = \frac{2\delta}{\pi\Delta}\left[\frac{1}{n} + 2n \sum_{p=1}^{\infty} (-1)^p \frac{(1 - \mathrm{e}^{-2p\pi/\Delta} \cos n\pi)}{(2p/\Delta)^2 + n^2}\right], \tag{3.187}$$

where the thickness $\Delta = \delta(x+1)/\pi$. If we take the limit $\delta \to 0$ with $x = O(1)$, we have

$$\begin{aligned} A_n(x,\delta) &\sim \frac{2\delta}{\pi\Delta}\left[\frac{1}{n} + 2n \sum_{p=1}^{\infty} \frac{(-1)^p}{(2p/\Delta)^2 + n^2}\right] \\ &= \delta \operatorname{cosech} \frac{n\delta(1+x)}{2} \end{aligned} \tag{3.188}$$

so that eq. (3.186) becomes Fay's solution

$$u_c = \delta \sum_{n=1}^{\infty} \frac{\sin n\tau}{\sinh \frac{n\delta}{2}(1+x)}. \tag{3.189}$$

The Fourier series representation of the sawtooth solution $(\pi - \tau)/(1+x)$ in $0 < \tau < 2\pi$ is

$$\frac{\pi - \tau}{1+x} = \sum_{n=1}^{\infty} \frac{\sin n\tau}{n}\left(\frac{2}{1+x}\right). \tag{3.190}$$

A comparison of eq. (3.189) with eq. (3.190) shows that, for larger n, the $1/n$ decay in the sawtooth solution has been replaced by $\mathrm{e}^{-n\delta(1+x)/2}$, displaying stronger dissipation due to nonlinearity as well as viscous diffusion. The solution (3.186) also shows that the shock thickness (with respect to τ) now is of order δx. Lesser and Crighton (1975) show that this solution holds even when $\delta x = O(1)$, and the viscous diffusion has spread from the narrow shock layer into the entire wave profile, a surprising result since it was derived as a Fourier series equivalent to a first order composite solution in which the shock thickness was assumed to be small, of order δ. But this is only representative of several other curious results that we have encountered. Thus, the first order composite perturbation solution (3.185) is itself an exact solution; its Fourier series representation, with further approximations in the Fourier coefficients, is another exact solution!

It is now appropriate to summarise how a wave profile produced by a periodic signal at $x = 0$ evolves with distance. The signal steepens with respect to τ and is described by the Fubini solution (3.154) until the shock is formed at $x = 1$. The shock evolves and adjusts until it has a balanced Taylor structure. This stage is *not* described by any of the analytic solutions found so far. Thereafter, we have the composite solution (3.185) which holds for some distance until it is replaced by Fay's solution (3.189). Fay's solution holds until the signal becomes so weak that it is essentially linear or has attained its 'old age', persisting to an infinitely long distance. The final old-age solution follows from eq. (3.189) as

$$u_c \sim \delta e^{-\delta x/2} \sin \tau. \tag{3.191}$$

This solution holds for $x \gg \delta^{-1}$ and has only a single Fourier component which decays under viscous diffusion alone. Of course, the amplitude of this solution is much smaller than that for the profile which decays due to linear diffusion throughout its evolution starting with the same initial profile as for (3.191) (see Blackstock (1964)). For further discussion of the equivalence of these solutions and their interpretation as a periodic array of spreading shocks that appear not to interact as they interpenetrate, reference may be made to the paper by Parker (1980). Whitham (1984) has generalised Parker's solutions for the Burgers equation, expressed as a sum of shocks, to nonlinear dispersive equations such as the Korteweg–deVries equation. The periodic solution in the latter case is represented as a sum of solitons, confirming the analogy between shocks and solitons as they appear in the analysis.

Scott (1981a) has considered the harmonic piston problem for cylindrical and spherical sources, using the boundary value problem analogue of eq. (3.118) for the governing equation. Two small parameters ε and R_0 appear in the problem, the former representing a combination of diffusive and geometrical effects, while the latter is a combination of nonlinear and geometrical effects. He has defined various matching asymptotic domains in the ε–R_0 plane and in each case the space-time asymptotic structure is discussed, providing in some cases the analytic form of the leading term. Rudenko and Soluyan (1977) have, in a somewhat *ad hoc* manner, attempted to 'string together' inviscid and viscous solutions of the non-planar equations to obtain composite solutions which have forms similar to the exact solutions of the planar equation. The problem for the non-planar case remains yet to be solved in complete generality.

Murray (1973) considered the generalised Burgers system

$$v_t + 2vv_y - Uv = \varepsilon v_{yy}, \tag{3.192a}$$

$$U_t + \varepsilon U = 1 - \int_0^1 v^2 \, dy, \tag{3.192b}$$

$$v(0, t) = 0 = v(1, t), \quad v(y, 0) = f(y), \quad U(0) = U_0 \tag{3.193}$$

as a model for turbulence, where v is the velocity of turbulent motion, $U(t)$ the velocity of the primary motion and $\varepsilon = 1/\mathrm{Re}$ the inverse Reynolds number. He employed singular perturbation techniques to find periodic solutions. The results indicated that an ultimate steady turbulent state emerges and that the small disturbances grow into a single large domain of relatively smooth flow, accomplished by a vortex sheet in which strong vorticity is concentrated.

A related system

$$u_t + uu_x = (\mu(t) + \mu_0)u_{xx}, \tag{3.194}$$

$$\mu_t + \gamma\mu = \lambda \int_{-\infty}^{\infty} u^2 \, dx, \tag{3.195}$$

$$u(x, 0) = u_0(x), \quad \mu(0) = 0, \tag{3.196}$$

where λ, γ and μ_0 are positive constants, appears in plasma physics. The existence, uniqueness and regularity of the solutions were discussed by Penel and Brauner (1974). Some numerical results, with a triangular initial profile, were also presented for a variety of the parameters involved. The profile evolves into a smooth single hump which decays with time.

3.7 An alternative approach to non-planar N waves

Here, we give an alternative analytic approach (Sachdev, Tikekar & Nair, 1986) to the N wave solution for the equation

$$u_t + uu_x + \frac{ju}{2t} = \frac{\delta}{2}u_{xx} \quad j = 1, 2 \tag{3.197}$$

(see sec. 3.4), based on the heuristic arguments inspired by the solution

$$u = \frac{x/t}{1 + \left(\frac{t}{t_0}\right)^{1/2} e^{x^2/2\delta t}} \tag{2.100}$$

for the plane case, $j = 0$. The basic idea is that the solution should have the form of the exact inviscid solution

$$u = \begin{cases} \dfrac{x}{2t}, & j = 1, \quad (3.198) \\ \dfrac{x}{t\ln t}, & j = 2, \quad (3.199) \end{cases}$$

near $x = 0$ and the asymptotic form (3.142) for large t. For $j = 0$, the latter agrees with the asymptotic form of (2.100) for large t. It turns out that the lobe Reynolds number (see eq. (2.101) for definition) for $j = 1$ can be found exactly by this method except for the value of the asymptotic constant in eq. (3.142). This does not seem possible in a unique manner for the spherical case. Thus, the present work, in a sense, complements the study of Crighton and Scott (1979) who were able to give some analytic results for the spherical case, but not for the cylindrical case.

Motivated by the form (2.100), we transform eq. (3.197) by

$$u = (2\delta)^{1/2} v, \quad \xi = \frac{x}{(2\delta t)^{1/2}}, \quad T = t^{1/2}, \tag{3.200}$$

so that it becomes

$$T v_T + (2Tv - \xi) v_\xi + jv = \tfrac{1}{2} v_{\xi\xi}. \tag{3.201}$$

Now, we introduce the inverse function V by

$$v = \frac{\xi}{V} \tag{3.202}$$

and 'peel' off' the factor $x/t^{1/2}$ from u. We obtain an equation with $\eta = x^2/2\delta t = \xi^2$ and T as independent variables:

$$\begin{aligned} V(jV - TV_T) + (2T - V)(V - 2\eta V_\eta) \\ + 3VV_\eta + 2\eta V V_{\eta\eta} - 4\eta V_\eta^2 = 0. \end{aligned} \tag{3.203}$$

Before we proceed to analyse eq. (3.203), we note that the solution (2.100) for $j = 0$ can be written as

$$\begin{aligned} u = (2\delta)^{1/2} v = (2\delta)^{1/2} \frac{\xi}{T(1 + aT\,e^{\xi^2})} \\ \equiv (2\delta)^{1/2} \frac{\xi}{V(\xi, T)}, \end{aligned} \tag{3.204}$$

where

$$a=(t_0)^{-1/2}$$

and

$$V=T(1+aT\,\mathrm{e}^{\xi^2})=T(1+aT\,\mathrm{e}^{\eta})$$
$$=T+aT^2+aT^2\eta+aT^2\frac{\eta^2}{2!}+aT^2\frac{\eta^3}{3!}+\cdots. \tag{3.205}$$

We seek a representation of the solution of eq. (3.203) in the form

$$V=f_0(T)+f_1(T)\eta+f_2(T)\frac{\eta^2}{2!}+f_3(T)\frac{\eta^3}{3!}+\cdots$$
$$=\sum_{i=0}^{\infty}f_i(T)\frac{\eta^i}{i!}, \tag{3.206}$$

which combines the asymptotic behaviour (3.142) and the inviscid behaviour (3.198)–(3.199). For the plane case, $f_0(T)$ is simply the sum of the remnant of the inviscid solution, namely T, and of the asymptotic time behaviour, aT^2; all higher order $f_i(T)$ just correspond to the asymptotic time factor aT^2. For other geometries, the solution is more complicated. Substituting (3.206) into eq. (3.203) and equating the coefficients of various powers of η to zero, we have

$$3f_1+2T-Tf'_0+(j-1)f_0=0, \tag{3.207}$$
$$5f_0f_2+2jf_0f_1-f_1^2-T(f_0f'_1+f_1f'_0)-2Tf_1=0, \tag{3.208}$$
$$7f_0f_3+[2(j+1)f_0-Tf'_0-3f_1-6T]f_2-Tf_0f'_2$$
$$+2f_1[(j+1)f_1-Tf'_1]=0, \quad \text{etc.} \tag{3.209}$$

It is easily verified that, for $j=0$, the functions $f_0=T+aT^2, f_1=f_2=f_3=aT^2$ satisfy eqs. (3.207)–(3.209). We consider the cylindrical and spherical cases separately.

(a) Cylindrical symmetry

Combining the information from the inviscid solution (3.198) and the asymptotic solution (3.142), we look for solutions to eqs. (3.207)–(3.209) in the form

$$f_0=2T+b_2T^2+b_3T^3,$$
$$f_1=c_1T+c_2T^2+c_3T^3,$$
$$f_2=d_1T+d_2T^2+d_3T^3, \quad \text{etc.} \tag{3.210}$$

We find by substitution in eqs. (3.207)–(3.209) that $c_1 = d_1 = 0$ and $c_2 = \frac{2}{3}b_2$, $c_3 = b_3$, $b_2 = \pm 3b_3^{1/2}$, $d_3 = b_3$, $d_2 = \frac{4}{15}b_2$ etc. so that the functions f_0, f_1, f_2 become known in terms of the constant b_3 which itself may be obtained from the numerical solution. That would also complete the asymptotic behaviour (3.142). Choosing the positive sign for b_2, to avoid any singularity at a finite time, we have

$$f_0 = 2T + 3b_3^{1/2}T^2 + b_3T^3. \tag{3.211}$$

All $f_i, i \geqslant 1$, can now be found by differentiation and algebraic operations from eqs. (3.207)–(3.209) etc. and hence, in principle, the solution is analytically found.

The solution for $j = 1$ can be written as

$$u = \frac{x}{t^{1/2}} \frac{1}{f_0(t^{1/2}) + f_1(t^{1/2})\eta + f_2(t^{1/2})\frac{\eta^2}{2} + \cdots} \tag{3.212}$$

so that

$$\left.\frac{\partial u}{\partial x}\right|_{x=0} = \frac{1}{t^{1/2} f_0(t^{1/2})} \tag{3.213}$$

where $f_0(t^{1/2})$ is given by eq. (3.211). Integrating eq. (3.197) with respect to x from 0 to ∞ and defining, as before, the lobe Reynolds number $R = (1/\delta)\int_0^\infty u \, dx$, we get an equation for R:

$$\begin{aligned} \frac{dR}{dt} + \frac{R}{2t} &= -\tfrac{1}{2}u_x(0, t) \\ &= -\frac{1}{2t^{1/2}} \frac{1}{2t^{1/2} + 3b_3^{1/2}t + b_3t^{3/2}}. \end{aligned} \tag{3.214}$$

Eq. (3.214) can be integrated to obtain

$$R = \left(\frac{t_0}{t}\right)^{1/2} + \left(\frac{1}{tb_3}\right)^{1/2} \ln \frac{t^{1/2} + 2/b_3^{1/2}}{t^{1/2} + 1/b_3^{1/2}}. \tag{3.215}$$

Here t_0 is another constant. The two constants t_0 and b_3 are chosen to match eq. (3.215) with the numerical solution (see sec. 5.6).

(b) Spherical symmetry

In this case if we peel off T from f_i by writing

$$f_0 = TF_0, \quad f_1 = TF_1, \quad f_2 = TF_2 \text{ etc.} \tag{3.216}$$

and introduce the variable $\tau = \ln T$, we get an autonomous system:

$$3F_1 - \frac{\mathrm{d}F_0}{\mathrm{d}\tau} + 2 = 0,$$

$$5F_0F_2 - F_1(2+F_1) - F_0\frac{\mathrm{d}F_1}{\mathrm{d}\tau} - F_1\frac{\mathrm{d}F_0}{\mathrm{d}\tau} + 2F_0F_1 = 0, \quad \text{etc.} \tag{3.217}$$

We may now find F_i as a power series in τ, that is, $\ln T$; the functions F_1, F_2, etc. can be expressed in terms of F_0, but F_0 itself cannot be found explicitly. It appears that the solution for $j = 2$ is much more complicated, involving expansions in powers of both t and $\ln t$. This is borne out by the analytic results of Crighton and Scott (1979). Thus, the shock displacement involves the exponential integral $\mathrm{Ei}(x) = ⨍_{-\infty}^{x} t^{-1}\mathrm{e}^t\,\mathrm{d}t$ (see eq. (3.141)), while the other asymptotic solution found by Crighton and Scott (1979) in the shock regime contains erfc and exponential functions. Crighton and Scott note that the non-uniformity of the singular perturbation solution first appears only in the shock region so that the outer solution near the node of the lobe continues to hold for a long time. The difficulty in our analytic solution is that f_0(or F_0) is not uniquely determined. However, once f_0 is chosen, all $f_i (i \geqslant 1)$ are uniquely found. One plausible choice for $f_0(t)$ is $(t^{1/2}\ln t + at^2)$. This is the sum of the inviscid and asymptotic contributions as in the planar case, wherein a is to be obtained from the numerical solution. We shall return to a discussion of these results in sec. 5.6, when we compare numerical solutions with the analytic ones.

3.8 Generalised Burgers equations and their transformations

Since generalised Burgers equations are difficult to treat analytically, as we have seen, there have been several attempts to generalise the Hopf–Cole transformation so as to linearise more general nonlinear equations. Further, it is natural to postulate that a general linear parabolic equation with variable coefficients would result from a more general nonlinear equation of the Burgers type by a Hopf–Cole like transformation; one might start from the general linear parabolic equation and obtain general nonlinear equations equivalent to it by means of a nonlinear transformation. We shall summarise some of these attempts.

Cole (1951) noted that the three-dimensional analogue of the Burgers equation

$$\mathbf{q}_t + \mathbf{q}\cdot\nabla\mathbf{q} = \nu\nabla^2\mathbf{q}, \tag{3.218}$$

which represents the Navier–Stokes equations for an irrotational flow without the pressure term, may be reduced to the linear heat equation in three dimensions by the transformation

$$\mathbf{q} = -2\nu\nabla(\ln\theta), \tag{3.219}$$

namely

$$\theta_t = \nu\nabla^2\theta. \tag{3.220}$$

This argument was generalised by Chu (1965), who considered the system of n equations

$$\frac{\partial u_i}{\partial t} + F_j\frac{\partial u_i}{\partial x_j} = G_i\frac{\partial u_i}{\partial x_j}\frac{\partial u_i}{\partial x_j} + \delta\frac{\partial^2 u_i}{\partial x_j\partial x_j} + H_i R_i, \quad (i, j = 1, \ldots, n), \tag{3.221}$$

where the summation convention is adopted with respect to the index j but the index i is not summed. The functions F_i, G_i and H_i are at least twice continuously differentiable functions of u_i. The R_i are continuously differentiable functions of $t, x_1, x_2, \ldots, x_n$, and δ is a constant. Chu introduced the transformation

$$F_i(u_i) = -\frac{2\delta}{\theta}\frac{\partial\theta}{\partial x_i}, \tag{3.222}$$

and substituted derivatives with respect to t, x_j etc. into eq. (3.221). The resulting equation, after an integration with respect to x_i, reduces to the linear parabolic equation

$$\frac{\partial\theta}{\partial t} = \delta\frac{\partial^2\theta}{\partial x_j\partial x_j} + \left[C(t) + \frac{P}{2\delta}\right]\theta, \tag{3.223}$$

if the functions F_i, G_i and H_i can be derived from a generating function $f_i(u_i)$ through

$$F_i = \int^{u_i} f_i(r)\,\mathrm{d}r$$

$$G_i = \delta\frac{\mathrm{d}\ln f_i}{\mathrm{d}u_i},$$

$$H_i = f_i^{-1} \tag{3.224}$$

(where i is again not summed), and the function $P(x_i, t)$ is related to R_i by

$$R_i = -\frac{\partial P}{\partial x_i}. \tag{3.225}$$

The function $C(t)$ is an arbitrary function of t arising from the integration with respect to x_i of the transformed equation. We shall here discuss only a scalar equation rather than the system (3.221). Thus, once the generating function f is chosen, the original nonlinear equation can be constructed. For example, for $f = 0, 1, e^u, nu^{n-1}, \ln u, \ -\sin u$ and $\cos u$, we get the following variety of equations, respectively:

$$\frac{\partial u}{\partial t} = \delta \frac{\partial^2 u}{\partial x^2}, \tag{3.226a}$$

$$\frac{\partial u}{\partial t} + u\frac{\partial u}{\partial x} = \delta \frac{\partial^2 u}{\partial x^2} + R(x,t), \tag{3.226b}$$

$$\frac{\partial u}{\partial t} + e^u \frac{\partial u}{\partial x} = \delta \left(\frac{\partial u}{\partial x}\right)^2 + \delta \frac{\partial^2 u}{\partial x^2} + e^{-u} R(x,t), \tag{3.226c}$$

$$\frac{\partial u}{\partial t} + u^n \frac{\partial u}{\partial x} = \delta \frac{n(n-1)}{u}\left(\frac{\partial u}{\partial x}\right)^2 + \delta \frac{\partial^2 u}{\partial x^2} + \frac{R(x,t)}{nu^{n-1}}, \tag{3.226d}$$

$$\frac{\partial u}{\partial t} + u(\ln u - 1)\frac{\partial u}{\partial x} = \frac{\delta}{u \ln u}\left(\frac{\partial u}{\partial x}\right)^2 + \delta \frac{\partial^2 u}{\partial x^2} + \frac{R(x,t)}{\ln u}, \tag{3.226e}$$

$$\frac{\partial u}{\partial t} + \begin{pmatrix} \cos u \\ \sin u \end{pmatrix}\frac{\partial u}{\partial x} = -\delta \begin{pmatrix} \cot u \\ \tan u \end{pmatrix}\left(\frac{\partial u}{\partial x}\right)^2 + \delta \frac{\partial^2 u}{\partial x^2} + \begin{pmatrix} -\operatorname{cosec} u \\ \sec u \end{pmatrix} R(x,t), \tag{3.226f}$$

where $R(x, t)$ is any function integrable with respect to x. The transformation (3.222) merely requires that a positive solution of the linear parabolic equation (3.223), or its one-dimensional analogue, should exist; hence every solution of the original equation may be obtained from that of the corresponding linear parabolic equation. It is interesting to note that, even for $R = 0$, the heat equation provides solutions for a whole class of eqs. (3.226).

Rodin (1970) studied directly the second member of eqs. (3.226), namely

$$u_t + uu_x = \delta u_{xx} + R(x,t), \tag{3.226b}$$

which is an inhomogeneous Burgers equation and occurs in several applications (Karabutov & Rudenko, 1976, Crighton, 1979). He related this equation with a Riccati equation and found some special solutions (see sec. 3.10).

For the one-dimensional case ($j = 1$), eq. (3.223) can be rewritten by absorbing $C(t)$ into $P(x, t)$ and normalising the variables according to

$$\frac{x}{L} \to \bar{x}, \quad \frac{\delta t}{L^2} \to \bar{t}, \quad \frac{uL}{\delta} \to \bar{u}, \quad P(x,t) \to \frac{2\delta^2}{L^2} \bar{P}(\bar{x}, \bar{t}),$$

where L is a characteristic length of the wave phenomenon being considered. The transformed equation (wherein we drop the bar) is found to be

$$\frac{\partial \theta}{\partial t} = \frac{\partial^2 \theta}{\partial x^2} + P(x,t)\theta. \tag{3.227}$$

The transformation (3.222) becomes $u = -(2/\theta)\partial\theta/\partial x$, if $F(u) = u$. Eq. (3.227) is the canonical form of a general linear parabolic equation with variable coefficients. To see this, we consider the general equation

$$u_{xx} + a(x,t)u_x + b(x,t)u = c(x,t)u_t, \quad c(x,t) > 0. \tag{3.228}$$

The one-to-one analytic transformation

$$\xi = \int_0^x (c(s,t))^{1/2}\,\mathrm{d}s, \quad \tau = t, \tag{3.229}$$

takes eq. (3.228) to an equation of the same form but with $c(x, t) = 1$. Therefore we may assume $c(x, t) = 1$ in eq. (3.228) to begin with. If we now set

$$u(x,t) = \theta(x,t)\exp\left[-\frac{1}{2}\int_0^x a(s,t)\,\mathrm{d}s\right], \tag{3.230}$$

we arrive at an equation of the type (3.227). Hence eq. (3.227) represents a canonical form of linear parabolic equations with variable coefficients, without any loss of generality.

It is natural to expect that generalised Burgers equations should reduce to a general linear parabolic equation rather than the classical heat equation through the Hopf–Cole type transformations. We have now to examine how an initial value or initial boundary value problem for such a linear equation may be solved. In general, there are no explicit exact solutions. We shall summarise some of the important special approaches.

A first question that arises is whether there are any operators which map solutions of the heat equation onto solutions of linear parabolic equations. Indeed, Colton (1974) has constructed precisely the integral operator whose domain is the space of solutions to the heat equation rather than the space of analytic functions. In the context of the general analytic theory of partial differential equations, this approach is the natural one for parabolic equations; just as the integral operators for elliptic equations reduce, in the case of Laplace's equation, to taking the real part of an analytic function, in the case of the heat equation the integral operator of Colton reduces to the identity operator. Colton (1976) has given a method for approximating solutions to initial boundary value problems for eq. (3.227). The essential idea is to construct a family of solutions to eq. (3.227), analogous to the set of polynomial solutions to the heat equation, given by Rosenbloom and Widder (1959), namely

$$\begin{aligned} h_n(x,t) &= n! \sum_{k=0}^{[n/2]} \frac{x^{n-2k} t^k}{(n-2k)!\,k!} \\ &= (-t)^{n/2} H_n\left(\frac{x}{(-4t)^{1/2}}\right), \end{aligned} \tag{3.231}$$

where $H_n(z)$ denotes the Hermite polynomials. Widder (1962) showed that the set $\{h_n(x,t)\}$ was complete in the space of solutions to the heat equation which are analytic in the neighbourhood of the origin. Colton proved the completeness of the family of solutions to eq. (3.227), analogous to the family $\{h_n(x,t)\}$ for the heat equation. This was accomplished by using an integral operator approach (Colton, 1974). Subsequently Colton (1975) also gave an algorithm for this purpose and illustrated it with a numerical example.

While the above approach has several attractive theoretical and analytic features, the final solution has to be obtained by numerical techniques. To that extent it detracts from the usefulness of the method in finding analytic solutions of the generalised Burgers equations.

An alternative approach is to find exact fundamental solutions of the linear parabolic equations, which may then be used to find a general solution from an arbitrary initial distribution by a convolution of the initial distribution function and the fundamental solution. Swan (1977) has given several important references to previous work on the subject and has found explicit fundamental solutions for some linear parabolic equations, for a particular choice of the coefficients which now depend on the spatial variable only. For example, if the given equation is

$$\alpha_1(x)u_{xx} + \alpha_2(x)u_x + \alpha_3(x)u = u_t, \tag{3.232}$$

then

$$u(x,t) = \frac{A}{\theta(x)t^{1/2}} \exp\left\{ -\frac{1}{4c^2 t}[h^2(x) - 2mh(x)t + (m^2 - 4c^2\mu)t^2] \right\} \tag{3.233}$$

is its fundamental solution if

$$\left.\begin{aligned} h(x) &= \int^x \frac{d\xi}{\theta(\xi)}, \\ \alpha_1(x) &= c^2\theta^2(x), \quad \alpha_2(x) = 3c^2\theta(x)\theta'(x) - m\theta(x), \\ \alpha_3(x) &= c^2\{[\theta'(x)]^2 + \theta(x)\theta''(x)\} - m\theta'(x) + \mu, \end{aligned}\right\} \tag{3.234}$$

where $\theta(x)$ is an arbitrary twice continuously differentiable function, and c, m, μ and A are arbitrary real constants. A and c are non-zero. Swan (1977) has also considered a slightly more general equation which includes the important Fokker–Planck equation

$$c^2 \frac{\partial^2}{\partial x^2}[\theta^2(x)u] - \frac{\partial}{\partial x}\{[c^2\theta(x)\theta'(x) + m\theta(x)]u\} - \frac{\partial u}{\partial t} = 0 \tag{3.235}$$

as a special case (see also Rogers (1983)). Besides the two approaches referred to above, one may resort to similarity analysis of the linear parabolic equation to identify the similarity variables and hence find the special similarity solutions. The latter could then be transformed to the solutions of the corresponding generalised Burgers equations via Hopf–Cole like transformations. Humi (1977) considered the equation

$$A(x)u_{xx} + B(x)u_x + C(x)u = u_t \tag{3.236}$$

with spatially varying coefficients, subjected it to infinitesimal transformations and identified several similarity transformations. Lehnigk (1976a, b) considered the same equation and discussed the similarity solutions which are conservative, that is, for which $\int_{x^*}^{\infty} u(x,t)\,dx$ is a finite constant. Here x^* is a constant equal to 0 or $-\infty$.

Thus, there is a possibility of generating a whole variety of solutions through a combination of a similarity approach to the more general linear parabolic equation (3.227) and Hopf–Cole like transformations.

We now discuss a few examples of this type (Sachdev, 1976b).

(a) Similarity solutions

If we assume $\theta = t^n \psi(z)$, where $z = x^2/4t$ and n is a real number, then eq. (3.227) reduces to the ordinary differential equation

$$z\psi'' + (z + \tfrac{1}{2})\psi' + (Q(z) - n)\psi = 0, \tag{3.237}$$

provided

$$P(x, t) = \frac{1}{t} Q(z). \tag{3.238}$$

Eq. (3.237) can be easily treated. Several special cases are now considered.

(i) If $Q(z) = z/4$, eq. (3.237) admits the solution

$$\psi = \mathrm{e}^{-z/2}(C_1 + C_2 z^{1/2}), \tag{3.239}$$

where C_1 and C_2 are arbitrary constants. In particular, if $C_1 = 1, C_2 = 0$, we have $\theta = t^{-1/4}\mathrm{e}^{-z/2}$. The corresponding nonlinear solution $u = x/2t$ satisfies the equation

$$u_t + uu_x + \frac{x}{4t^2} = u_{xx} \tag{3.240}$$

and represents a sawtooth form (fig. 5 of Benton and Platzman (1972) (B & P)). The solution $\theta = t^{-1/4} z^{1/2} \mathrm{e}^{-z/2}$ arising from eq. (3.239) with $C_1 = 0$, $C_2 = 1$, gives $u = x/2t - 2/x$ and corresponds to (3.5) of B & P.

(ii) Rodin (1969) has considered the special case $n = 0$ of eq. (3.237) with

$$Q(z) = -(a_1 z^{1/2} + a_2 z), \tag{3.241}$$

where a_1 and a_2 are arbitrary constants (in Rodin's notation, $z = x^2/t$). Eq. (3.237) can be transformed into

$$\frac{\mathrm{d}^2\bar{\psi}}{\mathrm{d}Y^2} + \frac{1}{2(1+2\alpha)} Y \frac{\mathrm{d}\bar{\psi}}{\mathrm{d}Y} + \frac{\beta^2 + 2\alpha}{4(2\alpha+1)^2}\bar{\psi} = 0, \tag{3.242}$$

where

$$\bar{\psi} = \mathrm{e}^{-(\alpha z + \beta z^{1/2})}\psi, \quad Y = 2(2\alpha + 1)z^{1/2} + 2\beta,$$

$$\alpha = \frac{-1 \pm (1 + 4a_2)^{1/2}}{2}, \quad \beta = \pm \frac{2a_1}{(1 + 4a_2)^{1/2}}.$$

The erf solutions of eq. (3.242) are possible only if $\beta^2 + 2\alpha = 0$. These correspond to Rodin's solution and to (3.1) and (3.3) of B & P.

(iii) For general n, eq. (3.237), with Q the same as in eq. (3.241), can be changed to the form

$$Y_1\frac{\mathrm{d}^2\bar{\psi}}{\mathrm{d}Y_1^2}+(\tfrac{1}{2}-Y_1)\frac{\mathrm{d}\bar{\psi}}{\mathrm{d}Y_1}-\frac{\beta^2+2\alpha-4n}{4(1+2\alpha)}\bar{\psi}=0, \tag{3.243}$$

where

$$Y_1=\frac{-1}{4(1+2\alpha)}[2(1+2\alpha)z^{1/2}+2\beta]^2.$$

Eq. (3.243) is a confluent hypergeometric equation. It has solutions in terms of Hermite functions, so that

$$\theta=t^n\,\mathrm{e}^{\alpha z+\beta z^{1/2}}H_\nu[2(2\alpha+1)z^{1/2}+2\beta] \tag{3.244}$$

if

$$\alpha=-\tfrac{5}{8},\quad \beta=-8a_1,\quad \nu=2(64a_1^2-4n-\tfrac{5}{4}). \tag{3.245}$$

The constant a_1 can be chosen so as to make ν an integer. The solution (3.244) gives

$$u=\frac{x}{t}\left[\tfrac{5}{8}+4a_1z^{-1/2}+\tfrac{1}{4}z^{-1/2}\frac{H'_\nu\left(-\dfrac{z^{1/2}}{2}-16a_1\right)}{H_\nu\left(-\dfrac{z^{1/2}}{2}-16a_1\right)}\right], \tag{3.246}$$

corresponding to (3.4) of B & P.

(iv) The inhomogeneous Burgers equation (3.226b) directly admits a similarity solution

$$u=t^{-1/2}(G+\xi) \tag{3.247}$$

if

$$R(x,t)=t^{-3/2}\bar{R}(\xi),\quad \xi=z^{1/2}=\frac{x}{2t^{1/2}}.$$

Eq. (3.226b) becomes

$$\frac{\mathrm{d}^2G}{\mathrm{d}\xi^2}-2G\frac{\mathrm{d}G}{\mathrm{d}\xi}+2\xi+2\bar{R}(\xi)=0. \tag{3.248}$$

An integration of this equation leads to the standard Riccati form

$$\frac{\mathrm{d}G}{\mathrm{d}\xi^2}-G^2=-2\int[\xi+\bar{R}(\xi)]\mathrm{d}\xi. \tag{3.249}$$

Solutions of this equation for various choices of $\bar{R}(\xi)$ correspond to the single hump solution (figs. 9–11 of B & P).

(b) Product solutions

The product solution $\theta = T(t)X(x)$ of eq. (3.227) exists only if P is a function either of x or of t alone. In the latter case, in view of eq. (3.225), we arrive at the homogeneous form of the generalised Burgers equations (3.226) with $R = 0$ (but see remarks following eqs. (3.226)). Therefore, we consider the case $P = P(x)$. Taking $T = \mathrm{e}^{\lambda t}$, λ a constant, we get, from eq. (3.227),

$$X''(x) + [P(x) - \lambda]X = 0. \tag{3.250}$$

This equation is amenable to a very general discussion. For special $P(x)$, say, $P = bx, b$ a constant, eq. (3.227) has via (3.250) the solution

$$\theta = \mathrm{e}^{\lambda t} z^{1/2} J_{\pm 1/3}(\tfrac{2}{3} z^{3/2}), \quad z = b^{-2/3}(bx - \lambda), \tag{3.251}$$

which, for $b = -1$, becomes

$$\theta = \mathrm{e}^{\lambda t} \begin{Bmatrix} \mathrm{Ai} \\ \mathrm{Bi} \end{Bmatrix} (\lambda + x), \tag{3.252}$$

where Ai and Bi are the Airy functions. These solutions correspond to (1.1)–(1.6) of B & P (cf. also Crighton (1979)). For $P = bx^2$, eq. (3.250) has solutions expressible in terms of Weber–Hermite functions. We note that the families of solutions, expressible in terms of Bessel or Hermite functions, may be used to solve a general initial value problem (see Titchmarsh (1962), Grinberg (1969)). Such solutions would correspond to Table 4 of B & P. Finally, we note that since 'a non-zero constant' is not a solution of eq. (3.227), we cannot find 'allied' classes of solutions from present solutions by linear superposition with a constant (cf. B & P).

3.9 The inverse transformation

Now we can approach the problem in an inverse manner (Sachdev, 1978). We begin with the linear parabolic equation

$$\phi_t + b\phi_x + c\phi + f = \varepsilon a \phi_{xx}, \tag{3.253}$$

where a, b, c and f are functions of x and t, and enquire what nonlinear equations it will generate through the transformation

$$F(u) = k(x, t)(\ln \phi)_x. \tag{3.254}$$

This transformation is slightly more general than Chu's (1965) and introduces a variable function $k(x, t)$ in the Hopf–Cole transformation. We thus have

$$
\left.\begin{aligned}
k\frac{\phi_x}{\phi} &= F, \quad k\frac{\phi_{xx}}{\phi} = F'(u)u_x - \frac{k_x}{k}F + \frac{F^2}{k},\\
k\frac{\phi_t}{\phi} &= \varepsilon a\left(F'u_x - \frac{k_x}{k}F + \frac{F^2}{k}\right) - bF - ck - \frac{fk}{\phi},\\
k\frac{\phi_{xt}}{\phi} &= F'u_t + \varepsilon a\frac{FF'}{k}u_x + F\left(-\frac{k_t}{k}\right.\\
&\quad \left. + \frac{\varepsilon a}{k}\left(-\frac{k_x}{k}F + \frac{F^2}{k}\right) - \frac{b}{k}F - c - \frac{f}{\phi}\right),\\
k\frac{\phi_{xxx}}{\phi} &= \left(\frac{F}{k} - \frac{k_x}{k}\right)\left(F'u_x - \frac{k_x}{k}F + \frac{F^2}{k}\right)\\
&\quad + \left(F''u_x^2 + F'u_{xx} - \left(\frac{k_x}{k}\right)_x F\right.\\
&\quad \left. - \frac{k_x}{k}F'u_x + \frac{2FF'}{k}u_x - \frac{F^2}{k^2}k_x\right).
\end{aligned}\right\} \tag{3.255}
$$

Differentiating eq. (3.253) with respect to x and rearranging, we have

$$
k\frac{\phi_{xt}}{\phi} + (b - \varepsilon a_x)k\frac{\phi_{xx}}{\phi} + (b_x + c)\frac{k\phi_x}{\phi} + kc_x + \frac{kf_x}{\phi} = \varepsilon ak\frac{\phi_{xxx}}{\phi}. \tag{3.256}
$$

Now we substitute (3.255) into eq. (3.256). To get an equation in u alone, we require that $f = 0$, that is, that eq. (3.253) is homogeneous. We thereby obtain

$$
\begin{aligned}
u_t &+ \left(b - \varepsilon a_x + 2\varepsilon a\frac{k_x}{k} - 2\varepsilon a\frac{F}{k}\right)u_x - \varepsilon a\frac{F''}{F'}u_x^2\\
&+ \frac{F}{F'}\left[\frac{\varepsilon a}{k^2}(kk_{xx} - 2k_x^2) - \frac{k_t}{k} - b\frac{k_x}{k} + \varepsilon a_x\frac{k_x}{k} + b_x + \frac{kc_x}{F}\right.\\
&+ \left. F\left(2\varepsilon a\frac{k_x}{k^2} - \varepsilon\frac{a_x}{k}\right)\right] = \varepsilon a u_{xx}.
\end{aligned} \tag{3.257}
$$

This equation is much more general than that with which Chu (1965) started his analysis. The term u_x^2 drops out if $F'' = 0$, that is, if F is a linear function of u. Further, eq. (3.257) can be put in a conservative form if the function $a(x, t)$ is a constant. Taking $F = u$, as a particular case, we get the following exchange process (cf. Murray (1968, 1970b)):

$$u_t + v_x + \left(\varepsilon a \frac{k_x}{k^2} u - \varepsilon a \frac{k_{xx}}{k} - \frac{k_t}{k} - \frac{bk_x}{k} \right) u + kc_x = 0,$$

$$v = \left(b + 2\varepsilon a \frac{k_x}{k} \right) u - \frac{\varepsilon a}{k} u^2 - \varepsilon a u_x. \tag{3.258}$$

This equation generalises the Burgers equation

$$u_t + v_x = 0, \quad v = -\varepsilon u_x + \frac{u^2}{2}, \tag{3.259}$$

and gives the class of nonlinear parabolic equations, with a constant coefficient of viscosity, which may be transformed into linear parabolic equations of the type (3.253).

We consider some other special cases of eq. (3.257). If we assume that the functions a and k are related as in the standard Hopf–Cole transformation, namely $k = -2a$, we can easily verify that eq. (3.257) becomes

$$u_t + (b + \varepsilon a_x + \varepsilon F) u_x - \varepsilon a \frac{F''}{F'} u_x^2 + \frac{F}{F'} \left[\varepsilon \left(a_{xx} - \frac{a_x^2}{a} \right) \right.$$

$$\left. - \varepsilon \frac{F}{2} \frac{a_x}{a} - \frac{a_t}{a} - \frac{ba_x}{a} + b_x - \frac{2ac_x}{F} \right] = \varepsilon a u_{xx}. \tag{3.260}$$

In particular, for $F = u$ this becomes the Burgers equation with damping and variable viscosity. The term $(b + \varepsilon a_x)$ in the coefficient of u_x, even if taken to be a constant, has an important role to play in the formation and decay of a shock (when $\varepsilon = 0$), as was shown by Murray (1970a). If we further specialise eq. (3.260) and choose $b + \varepsilon a_x = 0$, we have

$$u_t + \varepsilon u u_x - u \left(\frac{\varepsilon u a_x}{2a} + \frac{a_t}{a} \right) - 2ac_x = \varepsilon a u_{xx} \tag{3.261}$$

so that we have the possibility of having a damping term proportional to u^2, u or a constant, depending on the choice of the functions a and c. The variable function a on the right hand side can be viewed as a variable viscosity or simply a term meant to smooth the shock as in the viscosity method (cf. Dafermos (1973)). After choosing the form of a, we can choose other coefficients to have a variety of convective and damping terms. If, for example, $c = 0$, and a, and hence k, are functions of t alone, eq. (3.261) becomes

$$u_t + \varepsilon u u_x - \frac{a_t}{a} u = \varepsilon a(t) u_{xx}. \tag{3.262}$$

This includes the Burgers model for turbulence, considered by Case and Chiu (1969) and Murray (1973), but with a variable diffusivity (see also Romanova (1970) for its application to acoustic waves in the atmosphere).

Finally, if we assume a, k and c to be constant, but allow b to be a function of x and t, then eq. (3.260) becomes

$$u_t + (b + \varepsilon F)u_x - \varepsilon a \frac{F''}{F'} u_x^2 + \frac{F}{F'} b_x = \varepsilon a u_{xx}, \tag{3.263}$$

a useful equation with constant diffusivity, which for $F = u$ represents the Burgers type of equation with a linear damping term that may also depend on x and t through b_x.

Initial and nonlinear boundary value problems for a much more general Burgers hierarchy have recently been considered by Rogers and Sachdev (1984).

Nimmo and Crighton (1982) considered a wide class of Bäcklund transformations (BTs) for nonlinear parabolic equations of the form

$$u_t + u_{xx} + H(u_x, u, x, t) \equiv q + r + H(p, u, x, t) = 0,$$

with the usual notation

$$q = u_t, \quad p = u_x, \quad r = u_{xx} \text{ etc.},$$

where $H(u_x, u, x, t)$ is an arbitrary function of the indicated variables. These equations include both the Burgers equation with $H = 2uu_x$ and the inhomogeneous Burgers equation (3.266) (see sec. 3.10) with $H = 2uu_x - f(x, t)$ as special cases (after allowing for minor changes in the sign of t and some scaling). This paper also contains important references to the classical theory and applications of Bäcklund transformations with special reference to Korteweg–deVries and Burgers equations and their generalisations. Consider the transformation

$$p' = f(t, x, u, u', p, q), \quad q' = \psi(t, x, u, u', p, q), \tag{3.264}$$

with $p' = \partial u'/\partial x$ and $q' = \partial u'/\partial t$ as derivatives of the new dependent variable u'. We wish to find the functions f and ψ such that the pair of equations (3.264) is integrable if and only if u and u' satisfy the differential equations

$$\begin{aligned} r + q + H(p, u, x, t) &= 0, \\ r' + q' + G(p', u', x, t) &= 0. \end{aligned} \tag{3.265}$$

When this is the case, then eqs. (3.264) define a BT between the two equations (3.265).

Nimmo and Crighton (1982) arrive at the conclusion that the only nonlinear equations in the class considered that admit BTs are a slight generalisation of the Burgers equation, namely the inhomogeneous Burgers equations (3.266) (sec. 3.10). Further they claim that the only other class of equations that admit BTs of the kind (3.264) are linear parabolic equations with $H = a(x,t)u$. More specifically they show that BTs of the above type do not exist for the modified Burgers equation with cubic nonlinearity, for the cylindrical and spherical Burgers equations or, indeed, for any other equation of this class! This is in contrast to the equations of the KdV class where BTs exist for the KdV equation itself, for the modified KdV with cubic nonlinearity and for the cylindrical KdV equation.

Nimmo and Crighton (1982) have compared their analysis with Sachdev (1978) given in detail earlier in this section, and conclude that 'Sachdev's class of equations is no wider than the class we have obtained here', although it 'appears to contain equations that lie outside the class that we have shown to possess BTs'.

However, it should be noted that Bäcklund transformations of a reciprocal type which are not included in the class (3.264) in particular allow the linearisation of a hierarchy of nonlinear equations (see Rogers (1983), Kingston and Rogers (1982), Kingston, Rogers and Woodall (1984)). In particular, the nonlinear diffusion equation

$$\frac{\partial u}{\partial t} = \frac{\partial}{\partial x}\left[\frac{1}{(au+b)^2}\frac{\partial u}{\partial x}\right] + \frac{d}{(au+b)^2}\frac{\partial u}{\partial x}$$

may be linearised by a combination of reciprocal and Hopf–Cole type Bäcklund transformations. Applications of this result in two-phase flow in a porous reservoir have been given by Fokas and Yortsos (1982) and Rogers, Stallybrass and Clements (1983). Reference may be made to the text by Rogers and Shadwick (1982) for background material on Bäcklund transformations and to Rogers and Ames (1986) for other related matters.

3.10 The inhomogeneous Burgers equation

Besides the standard Burgers equation (2.44), the inhomogeneous Burgers equation

$$u_t + uu_x = \delta u_{xx} + f(x,t) \tag{3.266}$$

has attracted some attention as a one-dimensional scalar analogue of the Navier–Stokes equations. We have already referred to the work of Rodin

(1970) and Sachdev (1976b) in sec. 3.8. Here we summarise other work related to eq. (3.266). Kraut (1964) proved the uniqueness theorem for an initial boundary value problem for eq. (3.266) in the rectangle

$$0 \leqslant x \leqslant L, \quad 0 \leqslant t \leqslant T, \tag{3.267}$$

subject to the homogeneous boundary conditions

$$u(0,t) = u(L,t) = 0, \tag{3.268}$$

and the initial condition

$$u(x,0) = a(x), \tag{3.269}$$

where

$$a(0) = a(L) = 0, \tag{3.270}$$

without resort to the Hopf–Cole like transformation for linearisation. The proof requires that the coefficient of kinematic viscosity δ should be positive. (There is an interesting paper by Pelinovskii and Fridman (1974) wherein the consequences of taking δ to be negative have been discussed with reference to the Burgers equation.) Ton (1975) and Holland (1977) have shown that the solution of the homogeneous ($a(x) = 0$) initial boundary value problem (3.266)–(3.270) converges to the solution of the limiting equation as $\delta \to 0$; estimates of the rate of convergence were also given. Jeng and Meecham (1972) considered eq. (3.266) as a model equation for turbulence with the periodic driving force $f(x,t) = -A\sin\xi, \xi = kx - wt, A > 0$. The travelling wave solutions $u(x,t) = F(\xi)$ were found for the inviscid case $\delta = 0$. Periodic solutions arising from vanishing initial conditions were also found numerically, using a finite difference method. The solutions in the continuous part converge to the inviscid travelling wave solution. The shock is a discontinuity in the inviscid travelling wave solution, while it has a finite structure in the numerical solution depending on the Reynolds number of the wave (see also Okamura and Kawahara (1983)).

Karabutov and Rudenko (1976) proposed the inhomogeneous Burgers equation

$$\frac{\partial p}{\partial x} + \frac{1}{c_0}\frac{\partial p}{\partial t} - \alpha p \frac{\partial p}{\partial t} - \delta \frac{\partial^2 p}{\partial t^2} = \frac{c_0 \beta}{2c_p} Q(x,t) \tag{3.271}$$

to describe the uni-directional excitation and propagation of intense acoustic waves generated in a medium by the absorption of laser radiation. Here p is the incremental pressure in the medium, c_0 the speed of sound in the linear approximation, $\alpha = (\gamma + 1)/2\rho_0 c_0^3$ the nonlinearity parameter, δ the coefficient of high frequency viscous diffusion, and β the specific coefficient of volume expansion. The right hand side of eq. (3.271) represents absorption due to laser radiation. Notice the interchange of the roles of x and t, as in sec. 3.5.

The initial and boundary conditions relevant to eq. (3.271) are

$$p(x,0) = p(0,t) = 0. \tag{3.272}$$

The zero boundary condition at $x = 0$, referred to as 'rigid' by Karabutov and Rudenko, is found to lead to more effective excitation of waves.

The Hopf–Cole transformation

$$p = \frac{2\delta}{\alpha}\frac{u_t}{u} \tag{3.273}$$

or

$$u = \exp \int_0^t \frac{\alpha p}{2\delta} \mathrm{d}t$$

changes the problem (3.271)–(3.272) into

$$u_x + \frac{1}{c_0} u_t - \delta u_{tt} = u \frac{\alpha c_0 \beta}{4\delta c_p} \int_0^t Q(x,t)\, \mathrm{d}t, \tag{3.274}$$

$$u(x,0) = u(0,t) = 1, \quad x \geqslant 0, \quad t \geqslant 0. \tag{3.275}$$

The parameters of the medium determine the characteristic scales of the problem: $x_0 = 4\delta c_0^2$, $t_0 = 2\delta c_0$ and $p_0 = (\alpha c_0)^{-1}$. Changing the independent variables in eq. (3.274) to

$$z = \frac{x}{x_0}, \quad \tau = \frac{t}{t_0},$$

and denoting the non-dimensional combination of the parameters occurring in the problem by

$$A = \frac{\alpha c_0 \beta}{4\delta c_p} E_0, \tag{3.276}$$

where E_0 is the strength of the special short pulse

$$Q(x,t) = E_0 \delta(x) \delta(t), \tag{3.277}$$

we have

$$u_z + 2u_\tau - u_{\tau\tau} = A\delta(z)u. \tag{3.278}$$

This problem can be transformed by eliminating the first derivative term via $v = u\,\mathrm{e}^{z-\tau}$, so that we get the heat equation in v with the corresponding initial and boundary conditions:

$$v_z - v_{\tau\tau} = 0, \tag{3.279}$$

$$v(z,0) = e^{-A+z}, \quad v(0,\tau) = e^{-\tau}, \quad z \geqslant 0, \quad \tau \geqslant 0. \tag{3.280}$$

The solution of this problem may be found in Tychonov and Samarski (1964) and Copson (1975). We therefore have

$$u = e^{-(z-\tau)}v = \frac{1}{2}\left((1+e^{-A}) + (1-e^{-A})\left\{\phi\left(\frac{\tau-2z}{2z^{1/2}}\right)\right.\right.$$
$$\left.\left. - e^{2\tau}\left[1-\phi\left(\frac{\tau+2z}{2z^{1/2}}\right)\right]\right\}\right) \tag{3.281}$$

where $\phi(\xi) = (2/\sqrt{\pi})\int_0^{\xi} e^{-\alpha^2}\,d\alpha$ is the error function.

The solution for p is found with the help of eq. (3.273). While the solution (3.281) for u (and hence for p) is not a similarity solution, it arises from a special delta function form (3.277) of $Q(x,t)$ and represents an intermediate asymptotic (see chapter 4), since other solutions obtained on the basis of a finite scale Λ of the absorption region go over to this solution for $x \gg \Lambda$.

For $A \ll 1$, the first approximation to eq. (3.273) by way of eq. (3.281) is

$$\frac{p}{p_0} = A\left\{e^{2\tau}\left[\phi\left(\frac{\tau+2z}{2z^{1/2}}\right) - 1\right] + \frac{1}{(\pi z)^{1/2}}e^{-(\tau-2z)^2/4z}\right\}. \tag{3.282}$$

This is easily seen to be the solution of the linearised homogeneous form of eq. (3.271). In the second approximation when we retain linear terms in the expansion in A in eq. (3.281), we have

$$\frac{p}{p_0} = A\frac{\dfrac{1}{(\pi z)^{1/2}}e^{-(\tau-2z)^2/4z} + e^{2\tau}\left[\phi\left(\dfrac{\tau+2z}{2z^{1/2}}\right) - 1\right]}{1 + \dfrac{A}{2}\left[e^{2\tau}\left(\phi\left(\dfrac{\tau+2z}{2z^{1/2}}\right) - 1\right) + \phi\left(\dfrac{\tau-2z}{2z^{1/2}}\right)\right]}. \tag{3.283}$$

Eq. (3.283) shows more clearly how eq. (3.282) follows when A is very small. In the limit of $x \gg x_0$, that is, $z \gg 1$, eq. (3.283) becomes the single hump solution of the standard Burgers equation (see eq. (2.93)) so that, in this limit, the intermediate asymptotic nature of the solution becomes more apparent. It is also of importance to note that, under real experimental conditions, $x \gg x_0$, so that the solutions here have genuine physical significance. Figs. 3.1 and 3.2 give the height of acoustic pulse versus time at larger distances, and at specific finite points, respectively.

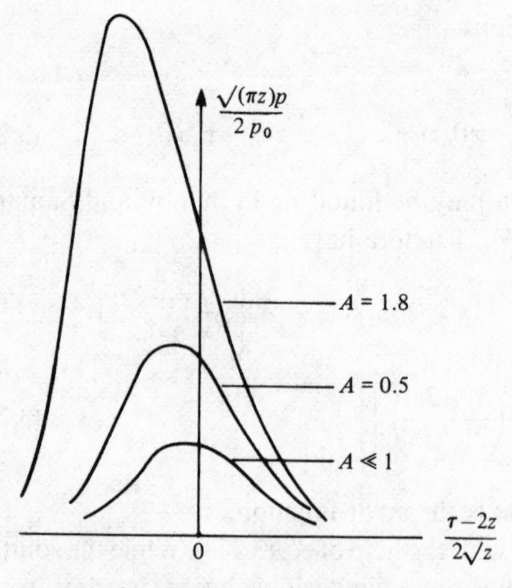

Fig. 3.1. Excitation of acoustic waves by laser radiation: height of acoustic pulse versus time at large distance for different source strengths: (1) $A = 1.8$; (2) $A = 0.5$; (3) $A \ll 1$. (From Karabutov and Rudenko (1976)).

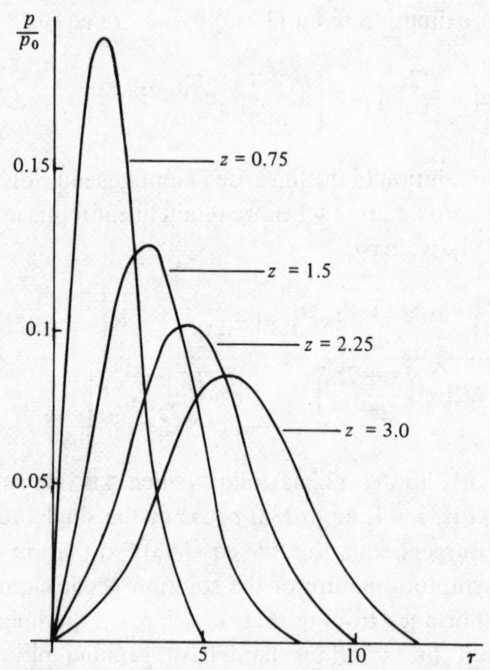

Fig. 3.2. Acoustic incremental pressure versus time at various point of medium for source strength $A = 0.5$ (from Karabutov and Rudenko (1976)).

On the other hand, as $A \to \infty$, eq. (3.281) has the limiting form

$$\lim_{A\to\infty} u \equiv L = \frac{1}{2}\left\{1+\phi\left(\frac{\tau-2z}{2z^{1/2}}\right)-e^{2\tau}\left[1-\phi\left(\frac{\tau+2z}{2z^{1/2}}\right)\right]\right\}, \tag{3.284}$$

so that

$$p_{\lim} = p_0 \frac{\partial L/\partial \tau}{L}. \tag{3.285}$$

In this case the pressure is finite for all $z \neq 0$ and $\tau \neq 0$, but becomes infinitely large when $\tau \to 0$. Fig. 3.3 gives pressure versus time at $z = 0.25$ for various values of A.

An equation more general than (3.271), which includes dispersion due to the influence of discrete structures of the medium etc., is likely to represent more realistically the surface absorption of laser radiation.

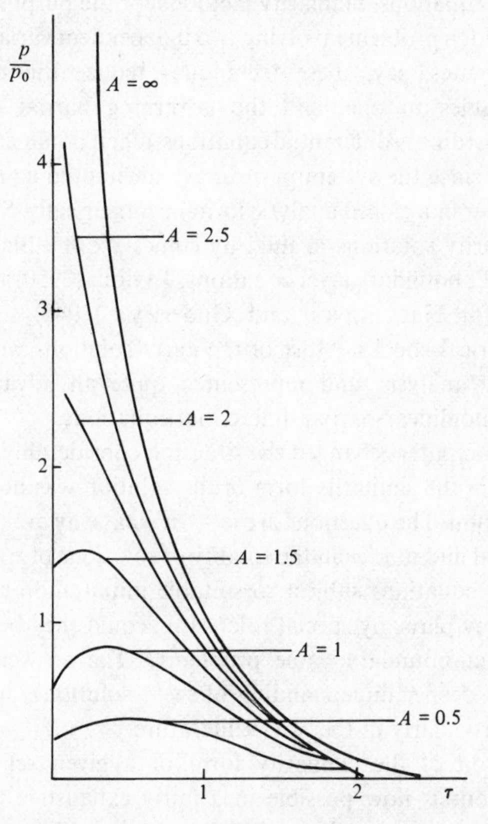

Fig. 3.3. Acoustic incremental pressure versus time at $z = 0.25$ for various absorbed energies (from Karabutov and Rudenko (1976)).

4 Self-similar solutions as intermediate asymptotics for nonlinear diffusion equations

4.1 The nature of self-similar solutions

Mathematicians have long attempted to find exact solutions of nonlinear partial differential equations. Similarity methods for the purpose are well-established. Thus, for a problem involving two independent variables, space and time co-ordinates, say, these techniques reduce the number of independent variables to one, and the governing partial differential equations become ordinary differential equations, albeit nonlinear. This is a considerable gain, since the system of ordinary differential equations can now be solved either in a closed analytic form or numerically. Some of the best known similarity solutions in fluid dynamics are the Blasius (1908) solution of Prandtl's boundary layer equations, Taylor's (1950) and Sedov's (1946) solutions for blast waves, and Guderley's (1942) solution for converging cylindrical shocks. Most of the early solutions were derived using dimensional analysis, and represented quite an advance in the understanding of nonlinear partial differential equations.

The advent of computers changed the situation considerably. Simplification accruing from the similarity form of the solution was not the most important motivation. The questions arose – 'In what way are the similarity solutions special and unique in the totality of solutions of a given set of partial differential equations subject to suitable initial and/or boundary conditions? Do they play any special role? How could they be related to more general initial/boundary value problems?' The answers to these questions led to a deeper understanding of these solutions, and to their classification – particularly in the Soviet literature.

The identification of the similarity form of a given set of partial differential equations is now possible in a fairly exhaustive manner by means of the invariance properties of the partial differential equations via the method of infinitesimal or finite transformations. However, these similarity forms have to be related to their initial and/or boundary conditions. It is here that one has to carefully distinguish their type.

Before proceeding, it is worthwhile to define self-similar (or automodel) solutions. A particular solution of a physical problem is called self-similar if the spatial distribution of the characteristics of motion, i.e., of the dependent variables, remains similar to itself at all times during the motion. In such circumstances, we can find time-dependent scales for all the dependent variables and the independent (spatial) variable, such that the dependent variables can be expressed as $u_i = m_i(t)F_i(x/l(t))$ (no summation with respect to i). This in turn leads to the definition of universal functions $u_i/m_i(t)$ which depend on the similarity variable $\xi = x/l(t)$ only.

One of the objectives of the present chapter is to show analytically, through several examples of nonlinear diffusion equations, that the similarity solutions describe 'intermediate asymptotic behaviour of wider classes of initial, boundary, and mixed problems; they describe the behaviour of these solutions away from the boundaries of the region of the independent variables. Alternatively, they represent solutions in a region where in a sense the solution is no longer dependent on the details of the initial and/or boundary conditions but is still far from being in a state of equilibrium' (Barenblatt & Zel′dovich, 1972). The solution may depend on the initial conditions in an integral form – representing the memory of the solution, derived from the initial conditions. Thus, as we shall see, a whole class of solutions arising from a certain class of initial conditions evolves asymptotically in time to the similarity solution of the given system of partial differential equations. We shall study through examples the nature of these initial/boundary conditions and the manner, both qualitative and quantitative, in which solutions tend to the similarity solutions. Alternatively, this may also be interpreted as a stability study of the similarity solutions with respect to certain types of changes in the initial conditions.

The similarity solutions may be categorised as belonging, in general, to two major classes – type I and type II. Solutions of type I are those which can be completely characterised by dimensional analysis of the problem. The independent variables and other dimensional parameters appearing in the problem fully determine the similarity variable $\xi = Art^{-\alpha}$: ξ is a non-dimensional variable so that the constant A has dimensions $\mathrm{L}^{-1}\mathrm{T}^{\alpha}$. For similarity solutions of the first type, the parameter α and the magnitude of A are obtained explicitly from the dimensional considerations of the physical problem. For the second type, this is not the case. In fact, even the exponent α is determined by a global study of the ordinary differential equations resulting from the similarity assumptions. These equations describe the internal structure of the waves. The solution of the equations is required to exist (in the large) subject to the boundary conditions of the problem. This constitutes an eigenvalue problem. The spectrum for the eigenvalue may be discrete or continuous. Further, stability considerations determine the unique value of the exponent α. The number A cannot be determined from the ordinary differential equations for the similarity solution. This can be found from the numerical solution of the entire system of partial differential equations subject to the class of appropriate initial conditions etc. such that the numerical solution matches the similarity solution. The matching determines the constant A. In this sense, the parameter A represents the memory of the non-self-similar régime of the solution and is an integral link between the self-similar and non-similar régimes. It is, in general, a complicated functional of the initial conditions of the problem.

One of the best known similarity solutions of the first kind is the Taylor–Sedov solution for blast waves. If the shock heading the blast is assumed to be strong so that the pressure ahead of it is zero, then the two parameters E_0, the energy of the blast wave, and ϱ_0, the density of the undisturbed medium, uniquely determine the similarity variable

$$\xi = \frac{r}{(E_0 t^2/\varrho_0)^{1/5}}.$$

Indeed, the exact solution can be written out. For diffusive equations, a simple example is provided by the idealised problem of heat conduction in an infinite medium due to the instantaneous supply of a finite amount of heat E at a given point, say, $x = 0$. Then, if c is the specific heat of the medium and κ its thermal diffusivity, both assumed to be constant, one may immediately write the similarity solution of the heat equation

$$u_t = \kappa u_{xx} \tag{4.1}$$

as

$$u = \frac{E}{c(4\pi\kappa t)^{3/2}} e^{-x^2/4\kappa t}. \tag{4.2}$$

The best-known example of similarity solutions of the second kind is the solution for a converging cylindrical shock (Guderley, 1942). For diffusive phenomena, we shall consider in detail the progressive wave solutions of Fisher's equation

$$u_\tau = \nu u_{yy} + ku(1-u), \tag{4.3}$$

a nonlinear partial differential equation of diffusion type which describes the propagation of a virile mutant in an infinitely long habitat. Here v and k are positive constants. It was first studied by Fisher (1936) and by Kolmogoroff, Petrovsky and Piscounov (1937) (KPP). The progressive wave solutions

$$u = U(y - c\tau + a), \tag{4.4}$$

with c and a constants, can be rewritten as

$$u = U(\ln \xi + c \ln T + a) = U_1\left(\frac{\xi T^c}{A}\right), \text{ say,} \tag{4.5}$$

by a simple change of variables $\xi = \mathrm{e}^y$, $T = \mathrm{e}^{-\tau}$, $A = \mathrm{e}^{-a}$. The parameter c now becomes the familiar similarity exponent which may be found by global analysis of the ordinary differential equation resulting from eq. (4.3) by substitution of (4.4), and the relevant boundary conditions. It turns out that it represents a similarity solution of the second kind for which the eigenvalue c has a continuous spectrum. It may also be observed that the above transformation brings the progressive wave solutions into a more familiar similarity form. The similarity solutions, therefore, include travelling waves, product solutions as well as stationary (time-independent) solutions as special cases. These are candidates for intermediate asymptotics of certain problems. We shall, in the present chapter, have occasion to deal with each of these kinds of solutions. Travelling solutions of eq. (4.3) are some of the earliest examples of similarity solutions of the second kind.

In fact, the self-similar solutions of both kinds represent degenerate cases of the complete non-self-similar problem in the sense that they arise as limiting solutions, as all constant parameters entering the initial and boundary conditions and having the dimensions of the independent variables either vanish or become infinite. It is this degeneracy that is reflected in the reduction of the number of independent variables and in the emergence of ordinary differential equations. Thus, the similarity solutions, *per se*, correspond in general to singular initial conditions in the form of generalised functions or their combinations. The limiting process is regular for solutions of the first kind and irregular for the solutions of the second kind, as we shall now show. Thus, according to the π-theorem (Sedov, 1982), a relation between $n+1$ dimensional quantities $a, a_1, \ldots, a_n$,

$$a = f(a_1, \ldots, a_k, a_{k+1}, \ldots, a_n) \tag{4.6}$$

can be recast as

$$\pi = F(\pi_1, \ldots, \pi_{n-k}) \tag{4.7}$$

where the parameters $\pi, \pi_1, \ldots, \pi_{n-k}$ are dimensionless quantities of the forms

$$\pi = \frac{a}{a_1^{\alpha} \ldots a_k^{\kappa}}, \quad \pi_1 = \frac{a_{k+1}}{a_1^{\alpha_1} \ldots a_k^{\kappa_1}}, \ldots, \tag{4.8}$$

$$\pi_{n-k} = \frac{a_n}{a_1^{\alpha_{n-k}} \ldots a_k^{k_{n-k}}}.$$

In the above, it is assumed that the quantities $a_1, \ldots, a_k$ have independent dimensions, and among $a_1, \ldots, a_n$ there are no more than k quantities with independent dimensions. Here, a may be one of the unknown (dependent) functions and $a_1, \ldots, a_n$ may include r, t and the physical parameters of the problem. If one of the dimensionless parameters, π_i, say, is either small or large, the situation is idealised and a limiting process is carried out such that π_i tends to zero or to infinity. If, for fixed values of all other parameters, the function $F(\pi_1, \ldots, \pi_{n-k})$ has a finite limit as π_i tends to zero or infinity, we have a regular limiting process. In this case, the parameter π_i drops out of the problem, leaving no imprint whatever. A self-similar problem of the first kind results and the solution can be found by purely dimensional considerations; in short, we have a completely self-similar phenomenon. If, on the other hand, this is not so, and instead we have either

$$\lim \pi_i^{-\alpha} F = \phi(\pi_1, \ldots, \pi_{i-1}, \pi_{i+1}, \ldots, \pi_{n-k})$$
$$\text{as } \pi_i \to 0 \quad \text{or} \quad \pi_i \to \infty, \tag{4.9}$$

or

$$F = \pi_i^{\alpha} \phi((\pi_j / \pi_i^{\beta}), \pi_1, \ldots, \pi_{i-1}, \pi_{i+1}, \ldots, \pi_{j-1}, \pi_{j+1}, \ldots, \pi_{n-k}) + \cdots$$

for some π_i and π_j and some constants α and β, as π_i and $\pi_j \to 0$ (the limit being non-zero and finite), then a self-similar problem of the second kind results. In this case the parameter π_i does not disappear from the problem. The self-similar solution is now determined via the solution of an eigenvalue problem with the relevant boundary conditions. We refer to this phenomenon as incompletely self-similar. Further details of self-similarity, illustrated by a variety of examples, may be found in the books by Zel'dovich and Raizer (1966) and Barenblatt (1979) (see also Newman (1983)).

In the present chapter, we shall study several nonlinear diffusion equations, their self-similar solutions, and the role of these solutions as intermediate asymptotics. Some of these equations are different from those of Burgers type, and are distinguished mainly by the absence of convective terms. Accordingly, they provide illuminating contrast with the Burgers form.

4.2 Fisher's equation

As the first example of nonlinear diffusion equations that are not of Burgers type, we consider Fisher's equation

$$u_\tau = \nu u_{yy} + ku(1-u). \tag{4.3}$$

Here ν is the diffusion coefficient, k a positive multiplication factor, τ is time and y is distance. This equation describes nonlinear evolution of a population u in a one-dimensional habitat. It was introduced by Fisher (1936) to describe the propagation of a virile mutant in an infinitely long habitat. It also represents, with a minor variation, a model equation for the evolution of a neutron population in a nuclear reactor, where the domain is obviously finite (Canosa, 1969, 1973). Eq. (4.3) describes a balance between linear diffusion and nonlinear local multiplication and admits shock-like solutions. However, as we shall see, this equation differs from the Burgers equation in several important physical and mathematical aspects; this is one reason for our choice of this equation as a representative of equations in which there is no convective term. We shall give first a simple analysis of the salient features and solutions of this equation.

If we introduce the non-dimensional time and distance

$$t = k\tau, \quad x = (k/\nu)^{1/2} y, \tag{4.10}$$

eq. (4.3) becomes

$$u_t = u_{xx} + u(1-u). \tag{4.11}$$

The habitat can support only a certain maximum population u per unit length, which, for simplicity, we choose to be unity so that

$$0 \leqslant u(x,0) \leqslant 1, \quad -\infty < x < \infty. \tag{4.12}$$

We look for all solutions of eq. (4.11) subject to (4.12) such that all the x-derivatives of u tend to zero as $x \to \pm\infty$, and

$$\lim_{x\to-\infty} u(x,t) = 1, \quad \lim_{x\to+\infty} u(x,t) = 0, \quad t \geqslant 0. \tag{4.13}$$

We note that the end conditions $u = 1$ and $u = 0$ are solutions of eq. (4.11). Physically, the first condition states that the population is saturated at the left end while the second condition denotes zero occupancy of the habitat at the right end. KPP proved that, for all initial conditions of the type (4.12), the solution of eq. (4.11) is also bounded for all time with the same bounds as in (4.12) so that

$$0 \leqslant u(x,t) \leqslant 1, \quad -\infty < x < \infty, \quad t > 0.$$

Furthermore, it was proved that for the initial conditions

$$u(x,0) = \begin{cases} 1, & x < 0, \\ 0, & x > 0, \end{cases} \tag{4.14}$$

or

$$u(x,0) = \begin{cases} 1, & x < a, \\ f(x), & a < x < b, \\ 0, & x > b, \end{cases} \tag{4.15}$$

where the function $f(x)$ is arbitrary, and a and b are finite numbers, the initial solution evolves, as $t \to \infty$, into a shock-like travelling wave which satisfies the conditions (4.13), and propagates to the right with the 'minimum allowable' characteristic speed $c_{\min} = 2$. Since the basic eq. (4.11) is invariant under the transformation $x \to -x$, it suffices to consider right-moving waves only.

First we prove, by phase plane analysis, that eq. (4.11) admits travelling wave solutions

$$u(x,t) = u(x - ct + d) \equiv u(s), \tag{4.16}$$

c denoting the speed of propagation of the wave. The constant d (cf. eqs. (4.4)–(4.5)) will not play any role in the present discussion, so it may be chosen to be zero. The solution form (4.16) of eq. (4.11) with boundary conditions (4.13) constitutes a self-similar solution of the second kind with an infinity of possible wave speeds $c \geqslant 2$, as we now show. If we substitute (4.16) into eq. (4.11) and take cognizance of the boundary conditions (4.13), we have

$$\frac{\mathrm{d}^2 u}{\mathrm{d}s^2} + c\frac{\mathrm{d}u}{\mathrm{d}s} + u - u^2 = 0, \tag{4.17a}$$

$$u(-\infty) = 1, \quad u(+\infty) = 0. \tag{4.17b}$$

Eqs. (4.17) define a nonlinear eigenvalue problem in an infinite domain, with the wave speed c as the eigenvalue. In the phase plane $(\mathrm{d}u/\mathrm{d}s, u) \equiv (y, u)$, eq. (4.17a) becomes

$$\frac{\mathrm{d}y}{\mathrm{d}u} = \frac{u^2 - u - cy}{y} \tag{4.18}$$

with singular points (0,0) and (0,1). Its linearised form is

$$\frac{\mathrm{d}y}{\mathrm{d}u}=\frac{-u-cy}{y}. \tag{4.19}$$

In the notation of Birkhoff and Rota (1978, p. 124), the discriminant relevant to eq. (4.19) is $\Delta = c^2 - 4$, and q is positive, being equal to 1. Since u is always positive, being the population of the habitat, we would want the point (0,0) to be a stable node. A centre or focus would require the trajectory to enclose the origin and would lead to u assuming negative values. The discriminant Δ should, therefore, be positive so that $c \geqslant 2$. We have thus a continuous infinite spectrum of eigenvalues for the problem (4.17). The other singular point (1,0) is easily seen to be a saddle point and the trajectories for values of $c \geqslant 2$ are as shown in fig. 4.1. This result was proved by both KPP and Fisher.

Since c is always greater than 2, $\varepsilon = 1/c^2$ turns out to be a convenient small parameter to solve eq. (4.18) by a perturbation method. If we introduce a new dependent variable $\bar{y} = cy$, eq. (4.18) becomes

$$\varepsilon\frac{\mathrm{d}\bar{y}}{\mathrm{d}u}=\frac{u^2-u-\bar{y}}{\bar{y}}. \tag{4.20}$$

This appears to be a singular first order nonlinear ordinary differential equation, since ε multiplies the highest order derivative, but a straightforward perturbation expansion,

$$\bar{y}(u,\varepsilon) = g_0(u) + \varepsilon g_1(u) + \varepsilon^2 g_2(u) + \cdots, \tag{4.21}$$

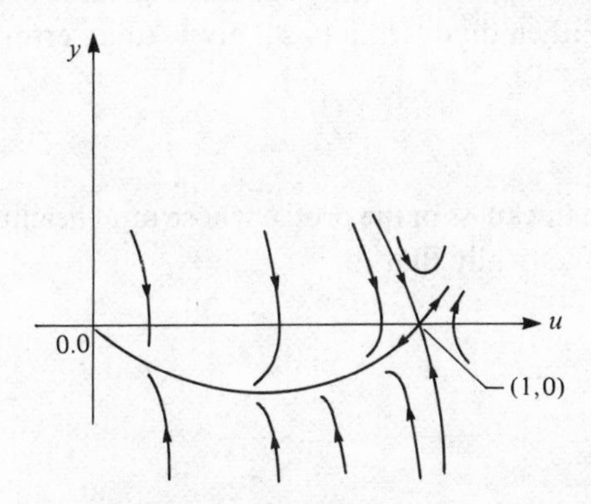

Fig. 4.1. Phase portrait of eq. (4.18): (0, 0) is a node and (1, 0) is a saddle point for $c \geqslant 2$. The integral curve joins these critical points. (From Canosa (1973).)

when substituted into eq. (4.20) gives

$$\begin{aligned} g_0 &= u^2 - u, \\ g_1 &= -g_0 g_0', \\ g_2 &= -g_0 g_1' - g_0' g_1, \quad \text{etc.} \end{aligned} \tag{4.22}$$

The solution passing through the singular points $u = 0$ and $u = 1$ comes out through mere substitutions; there is no need to solve the differential equations of various orders. Moreover, the boundary conditions are satisfied by terms of each order, as may easily be checked by inspection. The solution is

$$\begin{aligned} y(u, \varepsilon) = \varepsilon^{1/2}(u^2 - u) - \varepsilon^{3/2}(2u^3 - 3u^2 + u) \\ + 2\varepsilon^{5/2}(5u^4 - 10u^3 + 6u^2 - u) + O(\varepsilon^{7/2}). \end{aligned} \tag{4.23}$$

This 'singular' perturbation problem presents quite a contrast with the problems we have studied in chapter 3. The series (4.23) is an accurate asymptotic series as $\varepsilon \to 0\,(c^2 \to \infty)$. It is also a good approximation even when $\varepsilon = c^{-2} = 0.25$, corresponding to the slowest wave.

The solution (4.23) enables us to obtain a relation between shock wave thickness and its speed. The wave front is steepest at the point of inflexion, $\mathrm{d}^2y/\mathrm{d}s^2 = 0$, i.e., $y(\mathrm{d}y/\mathrm{d}u) = 0$. Assuming that $u = \frac{1}{2} + a\varepsilon$ at this point, substituting into $\mathrm{d}y/\mathrm{d}u = 0$ and using eq. (4.18), we get $a = -\frac{1}{4}$. Substitution into eq. (4.23) then gives the coordinates of the point of inflexion as

$$(u, y) = \left(u, \frac{\mathrm{d}u}{\mathrm{d}s}\right) = \left(\tfrac{1}{2} - \varepsilon/4, -(1 - \varepsilon^2/4)/4c\right). \tag{4.24}$$

If the steepness S of the wave profile is defined as the magnitude of the slope at its point of inflexion, then $\mathrm{d}u/\mathrm{d}s$ from (4.24) gives, to an error $O(\varepsilon^2)$,

$$S = \frac{1}{4c}. \tag{4.25}$$

Further, if L denotes the thickness of the profile whose total height is unity, it is easily checked geometrically that

$$L = \frac{1}{S} = 4c,$$

or

$$c = \frac{L}{4}. \tag{4.26}$$

This establishes an interesting result, namely, that the propagation speed of the wave is linearly proportional to its thickness; the time L/c for all the waves to pass a stationary observer is constant. In contrast to these results, the self-propagating shock governed by the Burgers equation has a unique speed provided by a conservation law and the end conditions. It is independent of the structure effected by the viscous diffusion; its thickness depends on its strength or speed in a much more complicated way (cf. eq. (2.87)). Besides, as we noted in chapter 2, the shock-like solution of Burgers equation is a self-similar solution of the first kind.

Before concluding the analysis in the phase plane, we note a curious relation between steady state solutions of the Korteweg–deVries–Burgers equation

$$h_t + hh_z + h_{zzz} = \delta h_{zz}. \tag{4.27}$$

(Johnson, 1970, 1972), and those of travelling waves of Fisher's equation. In eq. (4.27), δ is the coefficient of viscous damping. Introducing the variables

$$v(T) = \frac{1}{h_\infty} h(T), \quad T = \frac{z - \frac{1}{2}h_\infty t}{(2/h_\infty)^{1/2}}, \quad h(-\infty) = h_\infty, \tag{4.28}$$

into eq. (4.27), we get

$$\frac{\mathrm{d}^3 v}{\mathrm{d}T^3} + 2v\frac{\mathrm{d}v}{\mathrm{d}T} - \frac{\mathrm{d}v}{\mathrm{d}T} = \sigma \frac{\mathrm{d}^2 v}{\mathrm{d}T^2}, \tag{4.29}$$

$$\sigma = \delta\left(\frac{2}{h_\infty}\right)^{1/2}.$$

(Here h_∞ is a constant). An integration of eq. (4.29) and the use of undisturbed upstream and downstream boundary conditions, $v \to 0$, $T \to \infty, v \to 1, T \to -\infty$, give

$$\frac{\mathrm{d}^2 v}{\mathrm{d}T^2} + v(v-1) = \sigma \frac{\mathrm{d}v}{\mathrm{d}T}. \tag{4.30}$$

If we further change the variables according to

$$s = -T, \quad u = 1 - v,$$

we get precisely eq. (4.17a) with $\sigma = c$. The KPP–Fisher condition $\sigma = c \geqslant 2$ leads to the shock-like profiles of Johnson, which are monotonic both upstream and downstream. For $\sigma < 2$, steady solutions of eq. (4.27) are shocks which are oscillatory and damped downstream, and monotonic upstream. Johnson (1970) also carried out a perturbation analysis to confirm these results obtained by phase plane analysis and numerical integration. Zakharov and Korobeinikov (1980) have considered the invariant solutions of the nonplanar Korteweg–deVries–Burgers equation

$$u_t + \gamma u^m u_x + \frac{ju}{2t} - \mu u_{xx} + \beta u_{xxx} = 0$$

where $j = 0, 1, 2$ for plane, cylindrical and spherical symmetry, respectively, and m, γ, μ and β are constants. In several sub-cases corresponding to $j = 1$, $j = 2$, and for zero or non-zero values of γ, μ and β, the form of the solution was identified. Reference may be made to their original paper for physical applications of this generalised Korteweg–deVries–Burgers equation.

We now give the solution of eqs. (4.17) in the physical plane. Introducing the variable $z = s/c$ in eq. (4.17a) and using again the small parameter $\varepsilon = 1/c^2$, we have the boundary value problem

$$\varepsilon\frac{d^2u}{dz^2} + \frac{du}{dz} + u - u^2 = 0, \tag{4.31}$$

$$u(-\infty) = 1, \quad u(+\infty) = 0. \tag{4.32}$$

A regular perturbation series

$$u(z;\varepsilon) = u_0(z) + \varepsilon u_1(z) + \cdots \tag{4.33}$$

is assumed and substituted into eq. (4.31). The coefficients of various powers of ε are equated to zero to get

$$u_0' + u_0 - u_0^2 = 0, \tag{4.34}$$

$$u_1' + (1 - 2u_0)u_1 = -u_0'', \text{etc.} \tag{4.35}$$

If we choose the origin of z to be the point of inflexion of the profile, we have the initial conditions for u_i (following) from eq. (4.24) as

$$u_0(0) = \tfrac{1}{2}, \quad u_1(0) = -\tfrac{1}{4}, \text{etc.} \tag{4.36a, b}$$

Eqs. (4.34)–(4.35) subject to (4.36a, b), respectively, can be solved. The solution to $O(\varepsilon^2)$ is

$$u(s;\varepsilon) = \frac{1}{1 + e^{s/c}} - \varepsilon\frac{e^{s/c}}{(1 + e^{s/c})^2}\left[1 - \ln\frac{4\,e^{s/c}}{(1 + e^{s/c})^2}\right] + O(\varepsilon^2). \tag{4.37}$$

Canosa (1973) has compared the accuracy of this second order perturbation solution with the numerical solution of Fisher (1936) and the agreement is excellent over the entire profile thickness even for the smallest wave speed

$c = 2, \varepsilon = \frac{1}{4}$. The problem (4.31)–(4.32) is not really singular because, in the limit $\varepsilon \to 0, u \to 0$ or 1 and $du/dz \to 0$, eq. (4.31) has a uniform limit and the regular perturbation procedure gives an accurate solution in the entire domain. The solution (4.37) shows that the relaxation length or e-folding distance for the wave is c. Comparing this with eq. (4.26), we find that the profile thickness is four times the e-folding distance.

Since this solution represents an intermediate asymptotic (as KPP have proved analytically and Gazdag and Canosa (1974) have shown numerically), it is expected that it would be stable. The stability of nonlinear self-similar solutions has been discussed by Barenblatt (1979): a self-similar solution of a given problem is stable if the solution of any perturbed problem with sufficiently small perturbations can be represented in the form of a self-similar solution corresponding to a constant parameter that has, generally speaking, changed from its initial value, plus some additional term whose ratio to the unperturbed solution tends to zero as $t \to \infty$. Let $U = U(\zeta), \zeta = x - \lambda t + A$, be a self-similar solution which is perturbed at an initial time t_0 to

$$u(x, t_0) = U(x - \lambda t_0 + A) + \delta\phi(x) \equiv U(\zeta) + \delta\phi(x),$$

where δ is a small parameter and ϕ is non-zero only over a finite interval. Suppose the solution of this initial value problem, after a sufficiently long time, becomes

$$\begin{aligned} u(x, t) &= U(\zeta + a) + w(\zeta, t) \\ &\approx U(\zeta) + aU'(\zeta) + w(\zeta, t), \end{aligned}$$

where $w(\zeta, t) \to 0$ as $t \to \infty$ and where a is a small quantity corresponding to a small value of δ. In such a circumstance, the self-similar solution is stable, and, in fact, has evolved into another self-similar solution corresponding to a different value of A. The autonomous nonlinear equations of Fisher type, with the boundary conditions similar to (4.13), are invariant with respect to the two-parameter group of translational transformations

$$x' = x + \alpha, \quad t' = t + \beta, \quad u' = u.$$

In finding a solution of progressive type we seek a one-parameter subgroup of this transformation group corresponding to $\alpha = \lambda\beta +$ constant (where $\lambda = c$ is an eigenvalue), and a solution invariant with respect to that subgroup:

$$u(x', t') = u(x, t).$$

'The eigenvalues λ that extract from the basic group a one-parameter family are determined by the condition for the existence of an invariant solution in the large, i.e., the satisfaction of the boundary conditions by an invariant solution of the equation.' The progressive wave solution is thus determined up to a constant in the phase $\zeta = x - \lambda t + A$. Therefore, a definition of stability must have the corresponding invariance, and the perturbed solution should be allowed to tend to a solution, which is a shifted form of the original one. This shift does not spell instability.

However, if we admit perturbations of a non-finite extent, then the solution is not necessarily stable, as Canosa has shown for the case of Fisher's equation.

Now, we discuss the stability of eq. (4.11) by introducing a co-ordinate moving with the wave, $s = x - ct$, so that it becomes

$$u_t = u_{ss} + cu_s + u(1-u). \tag{4.38}$$

Let the perturbed solution be

$$u(s,t) = u(s;c) + \varepsilon v(s,t), \tag{4.39}$$

where the second term represents a small disturbance on the travelling wave with speed c. Substituting (4.39) in eq. (4.38), we have, to order ε,

$$v_t = v_{ss} + cv_s + [1 - 2u(s;c)]v. \tag{4.40}$$

We say that the solution $u(s;c)$ is stable if

$$\text{either} \lim_{t\to\infty} v(s;t) = 0 \text{ or } \lim_{t\to\infty} v(s,t) = u_s(s;c). \tag{4.41}$$

The first statement is the usual one for stability while the second implies that

$$u(s,t) = u(s;c) + \varepsilon u_s(s;c), \tag{4.42}$$

that is, the solution becomes

$$u(s - \delta s; t) \approx u(s;c) - \delta s u_s(s;c) \tag{4.43}$$

if $-\delta s = \varepsilon$; (4.43) represents a neighbouring travelling wave with a slight translation away from the original wave. That $u_s(s;c)$, the derivative of u with respect to s, is a (stationary) solution of eq. (4.40), is easily seen as follows. First, we put

$$v(s,t) = f(s)\,\mathrm{e}^{-\lambda t} \tag{4.44}$$

in eq. (4.40) to get the eigenvalue problem for $f(s)$:

$$f'' + cf' + (\lambda + 1 - 2u)f = 0, \quad f \to 0 \quad \text{as } s \to \pm\infty. \tag{4.45}$$

For $\lambda = 0$,

$$f(s) = u_s(s; c), \tag{4.46}$$

as is easily checked by differentiating eq. (4.38) with respect to s and setting $f = u_s(s; c)$. If the perturbation $v(s, t)$ is of finite extent, say, over $-L \leqslant x \leqslant L$, the eigenvalue problem (4.45) in terms of a new variable, $y(s) = \mathrm{e}^{cs/2} f(s)$, becomes

$$y'' + [\lambda - q(s)]y = 0, \quad y(\pm L) = 0, \tag{4.47}$$

where

$$q(s) = \frac{c^2}{4} - (1 - 2u) \geqslant 2u(s; c) > 0. \tag{4.48}$$

The last inequality follows from c being greater than or equal to 2. From standard results on eigenvalue problems (Birkhoff & Rota, 1978, p. 281) we conclude that all eigenvalues λ for (4.47) are real and positive and hence all perturbations of the wave having a finite extent decay exponentially with time.

However, consider an initial perturbation of the form

$$\varepsilon v(s, 0) = u(s; c + \delta c) - u(s; c), \tag{4.49}$$

leading to the evolution of $u(s; c + \delta c)$, a neighbouring travelling wave moving with a velocity δc in the frame of reference in which $u(s; c)$ is stationary. Then the solution $\varepsilon v(s, t)$ of eq. (4.40) with (4.49) as the initial condition is given by

$$\begin{aligned} \varepsilon v(s, t) &= u(s - (\delta c)t; c + \delta c) - u(s; c) \\ &= u(s; c + \delta c) - u_s(s; c + \delta c)(\delta c)t - u(s; c) \\ &= u(s; c) + u_c(s; c)\delta c - u_s(s; c + \delta c)(\delta c)t - u(s; c) \\ &= (\delta c)(u_c(s; c) - t u_s(s; c)), \end{aligned} \tag{4.50}$$

where a suffix denotes a partial derivative and terms of $O(\delta c)^2$ etc. are ignored. A direct substitution of (4.50) into eq. (4.40) confirms that $\varepsilon v(s; t)$ satisfies this equation. Eq. (4.50) means that, relative to an observer moving with the wave of velocity c, the faster of the two waves will move away from the other a distance which is linearly proportional to the time elapsed since the instant of the perturbation. Canosa has concluded that, in this sense, all travelling waves of Fisher's equation are unstable (cf. however Barenblatt (1979)). He has also demonstrated that 'superspeed' waves with $c > 2$ do not necessarily evolve into the minimum speed wave ($c = 2$) when subjected to arbitrary perturbations of infinite extent (see Hagstrom and Keller (1986)).

To supplement this analysis, Gazdag and Canosa (1974) have solved Fisher's equation numerically, using a pseudo-spectral approach for an accurate discretisation of the spatial derivatives (see sec. 5.5). They come to the following conclusions. (1) For a variety of initial conditions (including a step function and a wave with local perturbation), the solution evolves into the travelling wave of minimal speed $c = 2$. (2) For an initial super-speed $c > 2$, the evolution depends on the termination of the right hand tail of the wave. This is a necessary concomitant of their numerical method. Such truncation is physically plausible because the role of long distance dispersal in the spread of the genes is negligible. (3) Local perturbations are quickly smoothed even for superspeed waves. (4) From an initial distribution localised in space, two identical waves of minimum speed evolve, one propagating to the right and the other to the left.

Here, we may refer to the work of Larsen (1978) who has found upper and lower bounds for the solutions to the equations of Fisher type. These equations generalise eq. (4.11) so that $u(1 - u)$ is replaced by a function $F(u)$ possessing similar qualitative properties. Larsen has also considered the time-asymptotic (or intermediate asymptotic) convergence of transient solutions to travelling wave solutions, noting in particular the crucial dependence of long-time behaviour of the asymptotics on the manner of decay of the initial data at infinity.

We now discuss an exact solution of eq. (4.17) for a particular value of the wave speed c, and its relation to Painlevé transcendents (Ablowitz & Zeppetella, 1979). These functions are solutions of second order nonlinear differential equations whose only movable singularities are poles in the complex plane (no moving branch points or essential singularities). They were first classified by Painlevé and have been reviewed succinctly by Ince (1956). If we assume a singular solution of the type

$$u(s) \sim k(s - s_0)^{-\alpha} \tag{4.51}$$

for eq. (4.17a) where $\alpha > 0$, then the most singular contribution comes from $\mathrm{d}^2u/\mathrm{d}s^2$ and u^2 terms and these terms balance if $\alpha = 2$ and $k = 6$. Ignoring the translation constant s_0 corresponding to the movability of the pole, we seek a Laurent series solution

$$u(s) = 6/s^2 + a_{-1}/s + a_0 + a_1 s + \cdots . \tag{4.52}$$

Substitution of (4.52) in eq. (4.17a) determines a_{-1}, a_0, a_1, a_2 and a_3 uniquely as functions of c. However, the equation for a_4 has

$$0 \cdot a_4 + \frac{100}{40}\left(\frac{c}{5}\right)^2 - \frac{720}{8}\left(\frac{c}{5}\right)^6 = 0. \tag{4.53}$$

For this equation to be satisfied, either $c = 0$ or $c = \pm 5/\sqrt{6} \approx \pm 2.04$. The case $c = 0$ refers to stationary solutions, and is, in any case, integrable in closed form in terms of elliptic functions. The case $c = 5/\sqrt{6}$ belongs to the range of wave speeds discussed earlier. The solution in the neighbourhood of the origin, which, as we noted earlier, is a stable node, has the form (Lefschetz, 1977)

$$u(s) = \sum_{\substack{m,n \geqslant 0 \\ m+n \geqslant 1}} a_{mn}\, \mathrm{e}^{(m\lambda_1 + n\lambda_2)s}. \tag{4.54}$$

Substitution of (4.54) into eq. (4.17a) shows that, for $c = 5/\sqrt{6}, \lambda_1 = -2/\sqrt{6}$ and $\lambda_2 = -3/\sqrt{6}$ whence eq. (4.54) assumes the simpler form

$$u(s) = \sum_{n=2}^{\infty} a_n\, \mathrm{e}^{-ns/\sqrt{6}}, \tag{4.55}$$

where a_n satisfy the recurrence relation

$$a_n = \left[\frac{(n-2)(n-3)}{6}\right]^{-1} \sum_{j=2}^{n-2} a_j a_{n-j}, \quad n \geqslant 4, \tag{4.56}$$

for arbitrary a_2 and a_3. Since the range of s is $-\infty < s < \infty$, and the solution is required to be positive in the entire range including positive values of s, it follows that $a_2 > 0$. Choosing $a_2 = a^2$ and $a_3 = a^3 b_3$ leads to $a_n = a^n b_n$ in (4.56), where $b_2 = 1$ and

$$b_n = \frac{6 \sum_{j=2}^{n-2} b_j b_{n-j}}{(n-2)(n-3)}, \quad n \geqslant 4. \tag{4.57}$$

The solution (4.54) becomes

$$u = \sum_{n=2}^{\infty} a^n b_n\, \mathrm{e}^{-ns/\sqrt{6}} = \sum_{n=2}^{\infty} b_n \exp \frac{n(-s + \sqrt{6} \ln a)}{\sqrt{6}}. \tag{4.58}$$

Thus, a is merely a translation parameter and may be conveniently set equal to 1 without loss of generality. The free parameter b_3 identifies different trajectories moving into the node. The particular choice $b_3 = -2$ gives $b_4 = 3$ and, by induction from eq. (4.57), $b_n = (-1)^n(n-1)$. We are thus led to the compact form

$$u(s) = \sum_{n=2}^{\infty} (-1)^n (n-1)\, \mathrm{e}^{-ns/\sqrt{6}} = \frac{1}{(1 + \mathrm{e}^{s/\sqrt{6}})^2}. \tag{4.59}$$

This solution satisfies the boundary conditions $u(-\infty) = 1$ and $u(\infty) = 0$ and decreases monotonically. Other choices of b_3 lead to other such solutions. Ablowitz and Zeppetella have further shown that the transformation

$$u(s) = e^{\lambda_1 s} w(z),$$
$$z = \alpha\, e^{-(c+2\lambda_1)s} \tag{4.60}$$

converts eq. (4.17a) into

$$w'' = 6w^2 \tag{4.61}$$

for

$$\lambda_1 = -2/\sqrt{6}, \quad c = 5/\sqrt{6} \text{ and } \alpha = 1, \tag{4.62}$$

with solution $w = P(z-k; 0, g_3)$. Here, $P(s; g_2, g_3)$ is the Weierstrass P function with invariants g_2 and g_3 (Davis, 1962). In general, for $c = 5/\sqrt{6}$, eq. (4.17a) can be transformed into one of the 50 Painlevé type enumerated in Ince (1956). It is conjectured (see Ablowitz *et al.* (1980)) that nonlinear dispersive equations amenable to the inverse scattering method are all connected to Painlevé-type equations by appropriate similarity transformations.

Abdelkader (1982) has extended the above results to a generalised Fisher's equation. The travelling waves are governed by

$$u'' + cu' + u - u^n = 0, \quad 1 < n < \infty. \tag{4.63}$$

Abdelkader has shown that for

$$c = (n+3)/(2n+2)^{1/2}, \quad 2 < c < \infty,$$

the solution is expressible in terms of hyperelliptic integrals or, in some particular cases, elliptic integrals (see also Rosenau (1982) for a related equation describing a thermal wave in a reacting medium).

Finally, we refer to the work of Kametaka (1976) who has extended the results of KPP to an equation which generalises eq. (4.11) such that a function $f(u)$ with suitable properties replaces $u(1-u)$. Kametaka gives existence results for travelling waves, and stability theory for time-dependent as well as travelling wave solutions. His results are set in the context of earlier work of KPP.

4.3 A nonlinear heat equation

Now we study another nonlinear diffusion equation

$$\frac{\partial u}{\partial t} = \frac{\partial}{\partial x}[D(u)u_x], \tag{4.64}$$

which appears in a wide variety of physical applications, and has interesting analytic features. It coincides with the standard heat equation for $D(u) = 1$, which, under appropriate circumstances, is a good model for heat conduction in the absence of relaxation effects. However, when temperature variations are large, one must take into account a certain dependence of D on temperature. Eq. (4.64) has a rich history. *The mathematics of diffusion* by Crank (1975) devotes considerable attention to this equation; other texts which deal with this equation are Ames (1965) and Luikov (1966). If the nonlinear diffusion coefficient $D(u) = nu^{n-1}$ is such that $0 < n < 1$, $D(u) \to \infty$ as $u \to +0$, one refers to the phenomenon as fast diffusion, while, for $n > 1$, $D(u) \to 0$ as $u \to +0$ and the diffusion is called slow. While in heat conduction we generally deal with slow diffusion, there are circumstances in plasma physics (Bertsch, 1982) when, for example, $n = \frac{1}{2}$ so that the diffusion is fast.

A typical problem relevant to eq. (4.64) is an initial boundary value problem in a semi-infinite domain initially at a temperature (or concentration) u_0 when it is suddenly changed to u_1 at the left end boundary, say $x = 0$, and then held at that value thereafter. Another type of boundary condition consists of prescription of flux at $x = 0$. These formulations of the problem may be amenable to similarity form, so that the analysis becomes relatively simple. One may also pose a pure initial value problem for eq. (4.64) by prescribing $u(x, 0)$ initially with suitable behaviour at $x \to \pm\infty$. An interesting study of nonlinear diffusion equations of the type

$$\frac{\partial u}{\partial t} = x^{1-\lambda}\frac{\partial}{\partial x}(x^{\lambda-1}u^n u_x), \quad u \geqslant 0$$

is due to Grundy (1979). Here, λ is an index equal, in turn, to 1, 2 or 3 for plane, cylindrical and spherical geometries. Grundy considered the similarity solutions of this equation. He reduced the second order nonautonomous equation to a certain system of two first order equations, following an earlier work of Jones (1953). One of the members of the first order system is of the form

$$\frac{\mathrm{d}Y}{\mathrm{d}X} = \frac{F(X, Y)}{G(X, Y)}.$$

This reduction enables a study in the phase plane and a thorough discussion of the existence and uniqueness of all the solutions of the similarity type. Grundy has related his study to several previous investigations. We mention here another generalisation of eq. (4.64), namely

$$\frac{\partial u}{\partial t}=\frac{\partial}{\partial x}\left[D(u)\frac{\partial u}{\partial x}\right]+Q(u),$$

a nonlinear heat equation with a source term. Invariant solutions of this equation were found by Dorodnitsyn (1982).

Before we take up the study of initial and/or boundary value problems for eq. (4.64), we subject it to certain transformations in order to identify the similarity variable, as well as its canonical form (Munier *et al.*, 1981). This analysis illuminates the structure of the equation.

Choosing $D(u)=ku^n$ in eq. (4.64), we introduce the Kirchhoff transformation

$$v=\int_0^u t^n\,\mathrm{d}t.$$

The resulting equation in v is

$$\frac{\partial v}{\partial t}=k[(n+1)v]^{n/(n+1)}\frac{\partial^2 v}{\partial x^2}=Kv^{n/(n+1)}\frac{\partial^2 v}{\partial x^2},\quad\text{say.}\tag{4.65}$$

If we define the one-parameter continuous (multiplicative) group

$$\begin{aligned}v&=a^{\alpha}\bar{v},\\ t&=a^{\beta}\bar{t},\\ x&=a^{\gamma}\bar{x},\end{aligned}\tag{4.66}$$

where a is the parameter of the group and α, β and γ are constants to be determined such that eq. (4.65) is invariant under (4.66), we have

$$a^{\alpha-\beta}\frac{\partial\bar{v}}{\partial\bar{t}}=Ka^{[\alpha+(n/(n+1))\alpha-2\gamma]}\bar{v}^{n/(n+1)}\frac{\partial^2\bar{v}}{\partial\bar{x}^2}.\tag{4.67}$$

For invariance of eq. (4.65), we require that

$$\alpha-\beta=\alpha+\left(\frac{n}{n+1}\right)\alpha-2\gamma.\tag{4.68}$$

Since, from (4.66),

$$a=\left(\frac{v}{\bar{v}}\right)^{1/\alpha}=\left(\frac{t}{\bar{t}}\right)^{1/\beta}=\left(\frac{x}{\bar{x}}\right)^{1/\gamma},\tag{4.69}$$

the invariants of the group are

$$\eta=\frac{v}{t^{\alpha/\beta}}=\frac{\bar{v}}{\bar{t}^{\alpha/\beta}}\quad\text{and}\quad\xi=\frac{x}{t^{\gamma/\beta}}=\frac{\bar{x}}{\bar{t}^{\gamma/\beta}}.\tag{4.70}$$

Accordingly, eq. (4.65) (with $D(u) = ku^n$ in eq. (4.64)) admits the similarity form

$$v = t^{\alpha/\beta} f(x/t^{\gamma/\beta})$$
$$= t^{(n+1)(2\omega-1)/n} f(xt^{-\omega}), \tag{4.71}$$

where $\omega = \gamma/\beta$. Here we have used the condition (4.68). Substituting (4.71) into eq. (4.65), we get

$$\frac{n+1}{n}(2\omega - 1)f - \omega\xi f' = K f^{n/(n+1)} f''. \tag{4.72}$$

By choosing $\omega = 1/(n+2)$, this equation can be integrated once to give

$$f' = -k_0 \xi f^{1/(n+1)} + A,$$
$$k_0 = \frac{n+1}{K(n+2)}, \tag{4.73}$$

where A is a constant. The only case for which we have an explicit analytic solution for f corresponds to $A = 0$. In this case, we have

$$f^{n/(n+1)} = -\frac{n}{n+1} k_0 \frac{\xi^2}{2} + k_1, \tag{4.74}$$

where k_1 is an arbitrary constant. The solution for v becomes

$$v = t^{\alpha/\beta} f\left(\frac{x}{t^{\gamma/\beta}}\right) = \left[k_1 t^{-n/(n+2)} - \frac{1}{2}\frac{n}{n+1} k_0 \frac{x^2}{t}\right]^{(n+1)/n}, \tag{4.75}$$

leading to

$$u = (n+1)^{1/(n+1)} \left[k_1 t^{-n/(n+2)} - \frac{1}{2}\frac{n}{n+1} k_0 \frac{x^2}{t}\right]^{1/n}$$

or

$$u = \left[\tilde{K}(1+\mu t)^{-n/(n+2)} - \frac{1}{2}\frac{n}{n+2}\frac{\mu x^2}{1+\mu t} k_0\right]^{1/n}, \tag{4.76}$$

if an obvious translation and scaling in t are allowed. In the above, $\tilde{K}$ and μ are arbitrary constants. A strict similarity solution which does not decay or grow (explicitly) with time and is a function of the similarity variable $\eta = x/t^{1/2}$ only comes about when $\omega = \frac{1}{2}$; the parameter n remains arbitrary.

We next attempt to reduce eq. (4.64) to a nonlinear equation of the simplest type, exact linearisation such as that for the Burgers equation not being possible. For this purpose we seek Bäcklund transformations such that $u(x,t)$, x and t go, respectively, into

$$u(\hat{x},\hat{t}) = U\left(x,t,u,\frac{\partial u}{\partial x},\frac{\partial u}{\partial t}\right),$$

$$\hat{x} = X\left(x,t,u,\frac{\partial u}{\partial x},\frac{\partial u}{\partial t}\right), \tag{4.77}$$

$$\hat{t} = T\left(x,t,u,\frac{\partial u}{\partial x},\frac{\partial u}{\partial t}\right).$$

An extensive treatment of Bäcklund transformations may be found in the text by Rogers and Shadwick (1982) and of Lie–Bäcklund transformations in Anderson and Ibragimov (1979). It is again convenient to first introduce Kirchhoff's transformation for general $D(u)$, namely

$$\phi = F(u) = \int_0^u D(u)\,\mathrm{d}u.$$

We have, therefore,

$$\frac{\partial \phi}{\partial t} = F'(u)\frac{\partial u}{\partial t} = D(u)\frac{\partial u}{\partial t},$$

$$\frac{\partial \phi}{\partial x} = F'(u)\frac{\partial u}{\partial x} = D(u)\frac{\partial u}{\partial x}. \tag{4.78}$$

Assuming $F(u)$ to be a single valued function we may invert it so that

$$u = F^{-1}(\phi). \tag{4.79}$$

Equation (4.64), in view of (4.78) and (4.79), becomes

$$\frac{\partial \phi}{\partial t} = F'(u)\frac{\partial^2 \phi}{\partial x^2} = F'(F^{-1}(\phi))\frac{\partial^2 \phi}{\partial x^2} \equiv A(\phi)\frac{\partial^2 \phi}{\partial x^2}, \tag{4.80}$$

so that

$$A(\phi) = F'(F^{-1}(\phi)). \tag{4.81}$$

The next step is to simplify the nonlinearity $A(\phi)$ in eq. (4.80). A hodograph transformation is used such that the dependent and independent variables ϕ and x are interchanged:

$$\bar{\phi} = x, \quad \bar{x} = \phi; \quad \bar{t} = t. \tag{4.82}$$

Differentiating $\bar{\phi}$ with respect to t and x, we have

$$\frac{\partial \bar{\phi}}{\partial \bar{t}} + \frac{\partial \bar{\phi}}{\partial \bar{x}}\frac{\partial \phi}{\partial t} = 0, \quad \frac{\partial \bar{\phi}}{\partial \bar{x}}\frac{\partial \phi}{\partial x} = 1,$$

$$\frac{\partial^2 \bar{\phi}}{\partial \bar{x}^2}\left(\frac{\partial \phi}{\partial x}\right)^2 + \frac{\partial \bar{\phi}}{\partial \bar{x}}\frac{\partial^2 \phi}{\partial x^2} = 0. \tag{4.83}$$

Substituting (4.82) and (4.83) into eq. (4.80), we get

$$\left(\frac{\partial \bar{\phi}}{\partial \bar{x}}\right)^2 \frac{\partial \bar{\phi}}{\partial \bar{t}} = A(\bar{x})\frac{\partial^2 \bar{\phi}}{\partial \bar{x}^2} \tag{4.84}$$

The nonlinearity $A(\phi)$ has been rendered innocuous. The nonlinearity now appears as $(\partial\bar{\phi}/\partial\bar{x})^2$. However, eq. (4.84) suggests an obvious transformation

$$z = \left(\frac{\partial \bar{\phi}}{\partial \bar{x}}\right)^{-1}. \tag{4.85}$$

We differentiate eq. (4.84) with respect to $\bar{x}$ so that

$$\frac{\partial^2 \bar{\phi}}{\partial \bar{x}\partial \bar{t}} = -\frac{\partial}{\partial \bar{x}}\left[A(\bar{x})\frac{\partial}{\partial \bar{x}}\frac{1}{\partial\bar{\phi}/\partial\bar{x}}\right]. \tag{4.86}$$

In terms of z, this equation becomes

$$\frac{\partial z}{\partial \bar{t}} = z^2 \frac{\partial}{\partial \bar{x}}\left[A(\bar{x})\frac{\partial z}{\partial \bar{x}}\right] \tag{4.87}$$

which, when expressed in terms of the variables X and ζ, defined according to

$$\mathrm{d}X = \frac{\mathrm{d}\bar{x}}{A(\bar{x})}, \quad z(\bar{x}, \bar{t}) = \zeta(X, \bar{t}),$$

assumes the form

$$\frac{\partial \zeta}{\partial \bar{t}} = \zeta^2 a(X)\frac{\partial^2 \zeta}{\partial X^2}, \tag{4.88}$$

where

$$a(X) = \frac{1}{A(\bar{x})}.$$

Collecting the various transformations we have the connection between eqs. (4.80) and (4.87). That is, the Bäcklund transformation

$$\left.\begin{matrix} u(x,t) \\ x \\ t \end{matrix}\right\} \xrightarrow{\tau} \left\{\begin{matrix} \hat{u}(\hat{x},\hat{t}) = \dfrac{\partial u}{\partial x}, \\ \hat{x} = u, \\ \hat{t} = t, \end{matrix}\right\} \tag{4.89}$$

changes the nonlinear heat equation

$$\frac{\partial u}{\partial t} = A(u)\frac{\partial^2 u}{\partial x^2} \tag{4.90}$$

into

$$\frac{\partial \hat{u}}{\partial \hat{t}} = \hat{u}^2 \frac{\partial}{\partial \hat{x}}\left(A(\hat{x}) \frac{\partial \hat{u}}{\partial \hat{x}} \right). \tag{4.91}$$

Eq. (4.91) represents a canonical form of the nonlinear heat equation (4.64) via (4.89) in the sense that any nonlinear equation of the form (4.64) is equivalent to eq. (4.91) with a quadratic nonlinearity in the coefficient.

In the process of reduction, we have introduced non-autonomy in eq. (4.91) since now the right side of eq. (4.91) has $A(\hat{x})$ multiplying the second derivative $\partial^2 \hat{u}/\partial \hat{x}^2$. Munier *et al.* (1981) introduced further changes of variables to convert eq. (4.88) into the form

$$\frac{\partial v}{\partial t} = v^2 \frac{\partial^2 v}{\partial x^2} + \alpha \frac{v^3}{x^2}. \tag{4.92}$$

This is not much simpler than eq. (4.88) or eq. (4.91). However, it enabled them to establish homology between nonlinear equations of the type (4.64) with different coefficients $D(u)$, say, $D_1(u) = u^{n_1}$ and $D_2(u) = u^{n_2}$, n_1 or $n_2 \neq 2$. Such equations can be changed into the form (4.92) with different values of the parameter α. The homologous equations have solutions which are deducible from each other by simple transformations.

4.4 Asymptotic behaviour of solutions – intermediate asymptotics

Much later than the work of KPP referred to in sec. 4.2, Serrin (1967) proved that the similarity solutions are limiting forms of more general solutions of systems of partial differential equations. He considered specifically the system of Prandtl's boundary layer equations and showed that the solutions subject to a considerable class of boundary conditions at the leading edge of the plate converge to the Falkner–Skan similarity solutions at large distances downstream. Subsequently, similar results were proved for the nonlinear heat equation (4.64) by Peletier (1970) and Kamenomostskaya (1973, 1978).

We now briefly outline the procedure. First, the existence and uniqueness of the similarity solutions, governed by nonlinear ordinary differential equations, subject to physically motivated boundary conditions, are proved. The methods employed are often constructive. The solutions of the partial differential equations, with the same boundary conditions and with the asymptotic behaviour of their initial conditions the same as for the similarity solutions, are analysed, and their existence and uniqueness proved. Then it is proved that the solutions of the initial boundary value problems of the partial differential equations tend, as time tends to infinity, to the similarity solutions, pointwise and/or in the integral sense. The precise quantitative manner in which this convergence comes about is found as a consequence of the analysis.

In the following, we shall assume the existence and uniqueness of the solutions of the relevant ordinary and partial differential equations, and merely refer to the pertinent literature for further details.

Peletier (1970, 1971) has considered the asymptotic behaviour of the solution of the Cauchy–Dirichlet (initial boundary value) problem for eq. (4.64) in the half-strip $H_T = (0, \infty) \times [0, T]$ with initial condition $u(x, 0) = u_0(x), 0 \leqslant x < \infty$, and boundary condition $u(0, t) = U_0, t > 0$. The initial function $u_0(x)$ is required to satisfy the compatibility conditions $u_0(0) = U_0$ and $[D(u_0)u_0']' = 0$ at $x = 0$ and $u_0 \to B_0$ (a constant) as $x \to \infty$. The existence and uniqueness results for the similarity solution to this problem were obtained by Shampine (1973) and Atkinson and Peletier (1974), and the corresponding results for the partial differential equations were proved by Oleinik and Kruzhkov (1961). The smoothness requirements on $u(x, t)$ and $u_0(x)$ for the existence proof are assumed to be met, and the coefficient $D(u)$ is also assumed to be sufficiently smooth and positive for all $u > 0$.

For the purpose of illustration, we take up here the pure (Cauchy) initial value problem for eq. (4.64) (Van Duyn & Peletier, 1977a):

$$u_t = (D(u)u_x)_x, \tag{4.64}$$

$$u(x, 0) = u_0(x), \; -\infty < x < \infty, \tag{4.93}$$

in the strip $S_T = (-\infty, \infty) \times (0, T)$ where T is some fixed positive number; the function $u_0(x) \to A_0$ as $x \to -\infty$, and $u_0(x) \to B_0$ as $x \to \infty$, A_0 and B_0 being fixed but arbitrary real numbers. The initial function $u_0(x)$ and the diffusion coefficient $D(u) > \Delta > 0$ are chosen to belong to $C^{2+\alpha}(R)$, where R

denotes the real line and $0<\alpha<1$ so that the solution exists and is sufficiently smooth, that is, it belongs to $C^{2+\alpha}(\bar{S}_T)$. Δ is a real positive number.

Eq. (4.64) can be transformed into an ordinary differential equation in the similarity variable

$$\eta = \frac{x}{(t+1)^{1/2}}$$

so that $u(x,t) = f(\eta)$ satisfies

$$[D(f)f']' + \tfrac{1}{2}\eta f' = 0, \tag{4.94}$$

wherein the prime denotes differentiation with respect to η. In order to compare the solution of the Cauchy problem for eq. (4.64) with that of eq. (4.94) we impose the boundary conditions

$$f(-\infty) = A, \quad f(+\infty) = B. \tag{4.95}$$

Van Duyn and Peletier (1977b) proved that the problem (4.94)–(4.95) has a unique solution $f(\eta; A, B)$ which is monotonically decreasing if $A > B$ and monotonically increasing if $A < B$.

First we examine the behaviour of the similarity solutions as $\eta \to \infty$. To be specific, we assume that f decreases monotonically from A at $\eta = -\infty$ to B at $\eta = \infty$. If we set $p(\eta) = D(f)f'$, then eq. (4.94) becomes

$$p' + \tfrac{1}{2}\eta[D(f)]^{-1}p = 0. \tag{4.96}$$

Since $D(f) > 0$ for all f, and $f \to B$ as $\eta \to \infty$, the integral $\int_0^\infty |[D(f)]^{-1} - [D(B)]^{-1}|\,d\eta$ exists; therefore, by eq. (4.96), $p = O(e^{-\eta^2/4D(B)})$ as $\eta \to \infty$, and hence, by another integration,

$$f(\eta; A, B) - B = O(\operatorname{erfc}\{\eta/2[D(B)]^{1/2}\}) \quad \text{as } \eta \to \infty. \tag{4.97}$$

Similarly,

$$f(\eta; A, B) - A = O(\operatorname{erfc}\{-\eta/2[D(A)]^{1/2}\}) \quad \text{as } \eta \to -\infty. \tag{4.98}$$

As stated earlier, we impose the same behaviour on the initial function $u_0(x)$ by putting $\eta = x$ in (4.97)–(4.98):

$$u_0(x) - A_0 = O(\operatorname{erfc}\{-x/2[D(A_0)]^{1/2}\}) \quad \text{as } x \to -\infty,$$

$$u_0(x) - B_0 = O(\operatorname{erfc}\{x/2[D(B_0)]^{1/2}\}) \quad \text{as } x \to +\infty, \tag{4.99a, b}$$

where A_0 and B_0 are some specific values of the constants A and B.

Lemma 1. Let $u(x,t)$ be a solution of eqs. (4.64) and (4.93), with $u_0(x)$

satisfying (4.99). Then there exist numbers A^+, A^-, B^+, B^- such that

$$\max\{f(\eta; A_0, B^-), f(\eta; A^-, B_0)\} \leqslant u(x,t) \leqslant \min\{f(\eta; A_0, B^+), f(\eta; A^+, B_0)\} \text{ for } x \in R \text{ and } t \geqslant 0. \tag{4.100}$$

Proof. By (4.97),

$$f(x; A, B) - B = O(\operatorname{erfc}\{x/2[D(B)]^{1/2}\}) \quad \text{as } x \to \infty;$$

when $B = B_0$, $u_0(x)$ and $f(x; A, B_0)$ have the same asymptotic behaviour as $x \to \infty$. Moreover, $f(x; A, B_0) \to A$ and $u_0(x) \to A_0$ as $x \to -\infty$. Because of the monotonic behaviour of f, there exists a constant $A^+ > A_0$ such that

$$u_0(x) \leqslant f(x; A^+, B_0) \quad \text{for all } x \in R. \tag{4.101a}$$

Similarly there exists a constant $A^- < A_0$ such that

$$u_0(x) \geqslant f(x; A^-, B_0) \quad \text{for all } x \in R. \tag{4.101b}$$

It now follows from the maximum principle (Protter & Weinberger, 1967) that

$$f(x(t+1)^{-1/2}; A^-, B_0) \leqslant u(x,t) \leqslant f(x(t+1)^{-1/2}; A^+, B_0) \text{ in } \bar{S}_T. \tag{4.102}$$

Here we have kept the constant $B_0 = \lim_{x \to \infty} u_0(x)$ fixed. Using exactly the same argument, we can prove that

$$f(x(t+1)^{-1/2}; A_0, B^-) \leqslant u(x,t) \leqslant f(x(t+1)^{-1/2}; A_0, B^+) \text{ in } \bar{S}_T. \tag{4.103}$$

Combining (4.102) and (4.103) we get (4.100). We note that T is arbitrary. The inequality (4.100) provides preliminary bounds for the solution $u(x,t)$ in terms of similarity solutions. Now we proceed to obtain an estimate for the pointwise convergence of the solution $u(x,t)$ to the similarity solution.

Proposition. Let $A_0 > B_0$, and let there exist a positive number a such that

$$f(x+a; A_0, B_0) \leqslant u_0(x) \leqslant f(x-a; A_0, B_0) \quad \text{for all } x \in R.$$

Then the solution $u(x,t)$ of the initial value problem (4.64) and (4.93) converges towards $f(x(t+1)^{-1/2})$ as $t \to \infty$ and

$$\sup_{x \in R} |u(x,t) - f(x(t+1)^{-1/2})| \leqslant K(t+1)^{-1/2}, \quad t \geqslant 0,$$

where $K = 2a \max\{f'(\eta); \eta \in R\}$.

Proof. It follows from the hypothesis and the maximum principle that

$$f([x+a][t+1]^{-1/2}) \leqslant u(x,t) \leqslant f([x-a][t+1]^{-1/2}) \quad (4.104)$$

in $\bar{S}(T)$. Since

$$f\left(\frac{x}{[t+1]^{1/2}}\right) \geqslant f\left(\frac{x+a}{[t+1]^{1/2}}\right) \quad (4.105)$$

by the monotonicity of $f(\eta)$, reversing the sign in (4.105) and adding to the right inequality in (4.104) etc., we have

$$\begin{aligned}|u(x,t)-f(\eta)| &\leqslant |f([x-a][t+1]^{-1/2}) - f([x+a][t+1]^{-1/2})| \\ &\leqslant 2a(t+1)^{-1/2} \max\{f'(\eta); \eta \in R\} = K(t+1)^{-1/2} \quad \text{in } \bar{S}_T.\end{aligned} \quad (4.106)$$

Since the constant K is independent of T, the result follows.

Integral estimate

We now introduce η and $\tau = \ln(1+t)$ as the new independent variables in place of x and t. At $t=0$, $\eta = x$ and $\tau = 0$, so that the initial value problem (4.64) and (4.93) now becomes

$$u_\tau = [D(u)u_\eta]_\eta + \tfrac{1}{2}\eta u_\eta, \quad (4.107)$$

in the strip $(-\infty, \infty) \times (0, \ln[1+T])$, which may again be called S_T, and

$$u(\eta, 0) = u_0(\eta), \quad -\infty < \eta < \infty. \quad (4.108)$$

By Lemma 1, for any $\tau \in [0, \ln(1+T)]$,

$$\begin{aligned}|u-f| &= |(u-B_0)-(f-B_0)| \leqslant (u-B_0)+(f-B_0) \\ &\leqslant [\min\{f(\eta; A_0, B^+), f(\eta; A^+, B_0)\} - B_0] + (f-B_0).\end{aligned} \quad (4.109)$$

Therefore, $u-f$ is bounded and, in view of (4.97),

$$u(\eta,\tau) - f(\eta) = O(\operatorname{erfc}\{\eta/2[D(B_0)]^{1/2}\}) \quad \text{as } \eta \to \infty; \quad (4.110a)$$

similarly

$$u(\eta,\tau) - f(\eta) = O(\operatorname{erfc}\{-\eta/2[D(A_0)]^{1/2}\}) \quad \text{as } \eta \to -\infty \quad (4.110b)$$

for any $\tau \in [0, \ln(1+T)]$. Now, we let

$$\phi(\tau) = \int_{-\infty}^{\infty} [u(\eta, \tau) - f(\eta)] \, \mathrm{d}\eta. \tag{4.111}$$

In view of (4.110a, b), $\phi(\tau)$ is well-defined.

Lemma 2. The function $\phi(\tau)$ is given by

$$\phi(\tau) = \phi(0)\mathrm{e}^{-\tau/2}, \quad 0 \leqslant \tau < \infty. \tag{4.112}$$

Proof. Subtracting eq. (4.94) from eq. (4.107) and integrating with respect to η from $-\infty$ to $+\infty$, we have

$$\frac{\mathrm{d}\phi}{\mathrm{d}\tau} = [D(u)u_\eta - D(f)f_\eta]_{-\infty}^{\infty} + \tfrac{1}{2}[\eta(u-f)]_{-\infty}^{\infty} - \tfrac{1}{2}\phi(\tau). \tag{4.113}$$

From (4.110a, b) and the asymptotic behaviour of erfc, it is easily verified that $\eta(u-f) \to 0$ as $|\eta| \to \infty$. Since $f(\eta)$ tends to constant limits as $|\eta| \to \infty$, it follows that $f_\eta \to 0$ as $|\eta| \to \infty$. Again, (4.110a, b) and the fact that $u_0'(\eta) \to 0$ as $|\eta| \to \infty$ imply, by a standard argument, that u_η tends to zero as $|\eta| \to \infty$. Further, u(and $D(u)$) remain bounded; therefore, the first term on the right hand side of eq. (4.113) also vanishes. We arrive at the equation

$$\frac{\mathrm{d}\phi}{\mathrm{d}\tau} = -\tfrac{1}{2}\phi(\tau), \quad 0 \leqslant \tau \leqslant T, \tag{4.114}$$

which, on integration, gives eq. (4.112) if we allow T to tend to infinity.

Now we prove the main theorem.

Theorem 1. Let $u(x, t)$ be a solution of the problem (4.64) and (4.93), where u_0 satisfies the asymptotic conditions (4.99a, b), and let $\tilde{u}(\eta, t) = u(x, t)$. Then there exists a constant κ, which depends only on u_0 and f, such that

$$\int_{-\infty}^{\infty} |\tilde{u}(\eta, t) - f(\eta)| \, \mathrm{d}\eta \leqslant \kappa(t+1)^{-1/2}, \quad t \geqslant 0. \tag{4.115}$$

Proof. First we obtain an estimate for $|u - f|$. To that end, we construct two functions $u_0^+(\eta)$ and $u_0^-(\eta)$ such that they satisfy the same conditions as $u_0(\eta)$ and are chosen such that

$$u_0^+(\eta) \geqslant \max\{u_0(\eta), f(\eta)\}, \quad u_0^-(\eta) \leqslant \min\{u_0(\eta), f(\eta)\};$$

besides, $(u_0^+)_\eta$ and $(u_0^-)_\eta \to 0$ as $|\eta| \to \infty$. Let $u^+(\eta, \tau)$ and $u^-(\eta, \tau)$ be the solutions of eq. (4.107) assuming the initial conditions $u_0^+(\eta)$ and $u_0^-(\eta)$, respectively. Then, by the maximum principle,

$$u^+(\eta, \tau) \geqslant \max\{u(\eta, \tau), f(\eta)\},$$
$$u^-(\eta, \tau) \leqslant \min\{u(\eta, \tau), f(\eta)\}.$$

Therefore, when $u \geqslant f$,

$$|u - f| = u - f \leqslant u^+ - f$$

and, when $u \leqslant f$,

$$|u - f| = f - u \leqslant f - u^-.$$

Hence

$$|u - f| \leqslant (u^+ - f) + (f - u^-) \tag{4.116}$$

and

$$\int_{-\infty}^{\infty} |u - f| \mathrm{d}\eta \leqslant \int_{-\infty}^{\infty} (u^+ - f) \mathrm{d}\eta + \int_{-\infty}^{\infty} (f - u^-) \mathrm{d}\eta$$
$$\leqslant \mathrm{e}^{-\tau/2} \int_{-\infty}^{\infty} [u_0^+(\eta) - u_0^-(\eta)] \mathrm{d}\eta = \kappa(t+1)^{-1/2}. \tag{4.117}$$

The final step follows from the application of Lemma 2 to the solutions u^+ and u^-. The constant κ depends only on $u_0(\eta)$ and $f(\eta)$, and the integral it represents is well-defined due to the assumed asymptotic nature of $u_0^+(\eta)$ and $u_0^-(\eta)$.

Van Duyn and Peletier (1977a) have carried out further analysis to derive an improved regularity result in order to obtain a better rate of convergence for the integral estimate than (4.115).

Bertsch (1982) has extended the results of Van Duyn and Peletier (1977a) to the case $D(u) = mu^{m-1}, 0 < m < 1$. This case pertains to fast diffusion which occurs in plasma physics (see Bertsch for a physical discussion). While Bertsch has studied the Dirichlet problem in the semi-infinite domain $S_T = \{x, t{:}0 < x < \infty, 0 < t \leqslant T\}$, $T > 0$, Berryman (1977) and Berryman and Holland (1978) have discussed it in a finite domain for which separable solutions exist and serve as intermediate asymptotics. This problem is treated in the following section (see also Kamin and Rosenau (1981) and Rosenau and Kamin (1982) for similar results for thermal waves in inhomogeneous media).

4.5 A nonlinear diffusion problem arising in plasma physics

We generalise the nonlinear diffusion equation of sec. 4.3 and write

$$F(x)\frac{\partial n}{\partial t} = \frac{\partial}{\partial x}\left[D(n)\frac{\partial n}{\partial x}\right], \quad 0 \leqslant x \leqslant 1, \tag{4.118}$$

so that we have nonlinear diffusion coupled with a (space-dependent) geometrical factor $F(x)$. This equation describes particle diffusion across magnetic fields in a toroidal octupole plasma containment device (Drake *et al.*, 1977). Here n is the particle density, x is the spatial distance and t is the time. The geometrical factor $F(x)$ is a positive function describing the octupole geometry. The diffusion coefficient $D(n)$ is a nonlinear function of density, and, in the experiments of Drake *et al.*, it was found to follow Okuda–Dawson diffusion (Okuda & Dawson, 1972)

$$D(n) \propto n^{-1/2}. \tag{4.119}$$

In other density and field strength regions it would be given more generally by

$$D(n) \propto n^{\delta}, \tag{4.120}$$

with $\delta \geqslant -1$. In their experiment, Drake *et al.* observed the remarkable feature that, after a few milliseconds, the density profile evolved into a fixed shape, which then decayed in time. They referred to this time-independent (factor of the) density profile as the normal mode. This behaviour is suggestive of a separable solution of the relevant nonlinear equation (4.118) with suitable initial conditions. The experimental results on the toroidal octupole are well approximated by assuming null conditions on the boundaries:

$$n(0, t) = n(1, t) = 0. \tag{4.121}$$

The experimental results as well as the numerical studies (Berryman, 1977) indicated that the separable solution evolves from arbitrary initial data with boundary conditions (4.121) in a finite time. In this sense, the separable solution forms an intermediate asymptotic (Berryman, 1977, Berryman & Holland, 1978). With

$$D(n) = (1 + \delta)n^{\delta}, \quad \delta > -1, \tag{4.122}$$

(the case $\delta = -1$ would give a similarity solution requiring different boundary conditions), it is more convenient to introduce the pseudo-density

$$m(x,t) = n^{1+\delta}, \quad \delta > -1, \tag{4.123}$$

so that eq. (4.118) becomes

$$F(x)(m^{q-1})_t = m_{xx}, \tag{4.124}$$

where

$$q = \frac{2+\delta}{1+\delta}. \tag{4.125}$$

For $\delta > -1$, the function $m(x,t)$ has the same behaviour as $n(x,t)$ so that $0 \leqslant m(x,t) < \infty$ when $0 \leqslant n(x,t) < \infty$ and, furthermore, $m = 0$ when $n = 0$. The boundary conditions on m are

$$m(0,t) = m(1,t) = 0. \tag{4.126}$$

The geometrical factor $F(x)$ is positive in the cases of physical interest and may have an integrable singularity at an interior point $x = x_s$, say, of the square root or logarithmic type. Thus $F(x)$ satisfies

$$F(x) > 0 \quad \text{for } 0 \leqslant x \leqslant 1, \tag{4.127}$$

$$\int_0^1 F(x)\,\mathrm{d}x < \infty. \tag{4.128}$$

Moreover, it is consistent and convenient to have $F'(x) \geqslant 0$ for $x < x_s$ and $F'(x) \leqslant 0$ for $x > x_s$.

Eq. (4.124), which both is nonlinear and has a variable coefficient, is not solvable in a closed form. Therefore, first the analytic behaviour of the solution of eqs. (4.124)–(4.126) with $F(x) = 1$ will be studied, and then the influence of $F(x)$, both quantitative and qualitative, will be determined using analytic and numerical methods. In particular, the case $\delta = -\frac{1}{2}$, corresponding to $q = 3$, is of major physical interest and will be carefully analysed.

Taking $F(x) = 1$, eq. (4.124) becomes

$$(m^{q-1})_t = m_{xx}. \tag{4.129}$$

This equation has a quite general similarity solution (Ames, 1965). Restricting our consideration to the separable solution, we set

$$m = T(t)S(x). \tag{4.130}$$

Substituting (4.130) into eq. (4.129) we have

$$(q-1)T^{q-3}T' = S''S^{1-q} = -\lambda, \tag{4.131}$$

where λ is the separation constant. The solution for T is easily found to be

$$T = A_0\left[1 - \frac{\lambda(q-2)}{q-1}A_0^{-(q-2)}t\right]^{1/(q-2)} \tag{4.132}$$

where A_0 is an arbitrary constant.

According to eq. (4.131), the shape function $S(x)$ satisfies the nonlinear ordinary differential equation

$$\frac{d^2S}{dx^2} + \lambda S^{q-1} = 0. \tag{4.133}$$

This equation is invariant under the transformation

$$S(x) \to aS_0(x), \quad \lambda \to \lambda_0 a^{q-2}, \tag{4.134}$$

where a is a constant. If S_0 is a solution of eq. (4.133) with separation constant $\lambda = \lambda_0$, then $S(x) = aS_0(x)$ is also a solution with $\lambda = \lambda_0 a^{q-2}$. Therefore we can choose a such that $0 \leqslant S \leqslant 1$. We shall assume, without any loss of generality, that such a scaling has been performed and that S lies between 0 and 1. Multiplying eq. (4.133) by dS/dx and integrating, we get an implicit solution for S, namely

$$I(S) = \int_0^S \frac{dy}{(1-y^q)^{1/2}} = \varrho x, \quad 0 \leqslant x \leqslant \tfrac{1}{2}, \tag{4.135}$$

where $S = 0$ when $x = 0$ and

$$\varrho^2 = \frac{2\lambda}{q}. \tag{4.136}$$

Eq. (4.133), with homogeneous boundary conditions at $x = 0$ and $x = 1$, has a solution symmetric about $x = \frac{1}{2}$. The integral $I(S)$ can be expressed in terms of the incomplete Beta function or the hyper-geometric function as

$$I(S) = \frac{1}{q}B_S\left(\frac{1}{q}, \frac{1}{2}\right) = S^{1/q}{}_2F_1\left(\frac{1}{q}; 1 + \frac{1}{q}; S\right) \tag{4.137}$$

While one may find x for a given value of S between 0 and 1, it does not seem possible to invert eq. (4.135) in a simple way. We may, however, use it to find interesting relations between the various parameters of the problem. For example, if we put $S = 1, x = \frac{1}{2}$ in eq. (4.135), we have

$$\int_0^1 \frac{dy}{(1-y^q)^{1/2}} = \tfrac{1}{2}\varrho$$

or

$$\varrho = 2I(1) = \frac{2\,\Gamma(1/q)\Gamma(\frac{1}{2})}{q\,\Gamma(\frac{1}{2}+1/q)}, \tag{4.138}$$

as may easily be verified.

Eqs. (4.136) and (4.138) show that the eigenvalues

$$\lambda = \tfrac{1}{2} q \varrho^2 \tag{4.139}$$

become known for all admissible values of q between 1 and ∞ corresponding to the values of δ between $+\infty$ and $-1+0$ (see eq. (4.125)). Furthermore, the integral of the physical density's shape function is given by

$$\gamma \equiv \int_0^1 S^{q-1}(x)\,dx = 2\int_0^1 S^{q-1}(x)\frac{dx}{dS}dS$$
$$= (2/\rho)\int_0^1 \frac{S^{q-1}}{(1-S^q)^{1/2}}dS = \frac{4}{q\rho}. \tag{4.140}$$

The total number of particles, according to the separable solution (using eqs. (4.123), (4.125), (4.130) and (4.140)), is

$$N(t) = \int_0^1 n(x,t)\,dx = \gamma T^{q-1}(t) \tag{4.141}$$

for all time. It can be easily verified that, as δ continuously varies between $+\infty$ and $-1+0$, q varies between 1 and $+\infty$, ϱ varies between 4 and 2, γ between 1 and 0 and λ between 8 and $+\infty$, monotonically.

It is clear that, even when $F(x)$ is unity, no explicit analytic solution of eq. (4.118) of separable type can be found. Thus there is a need for an accurate numerical procedure to find the solution for general $F(x)$. A simple iterative method was given by Berryman (1977). The equation for $S(x)$, for the case of general $F(x)$, is easily found to be

$$-S''(x) = \lambda F(x) S^{q-1}(x). \tag{4.142}$$

If, for eq. (4.142), we set up an iterative process, with ith iterate $S_i = \Sigma_i$, obtained by solving the linear boundary value problem

$$\begin{aligned} &\Sigma_i''(x) = -\lambda F(x) S_{i-1}^{q-1}, \\ &\Sigma_i(0) = \Sigma_i(1) = 0, \end{aligned} \tag{4.143}$$

we can devise a suitable scaling so that this process converges; that is, $S_i \approx S_{i-1}$ after some i. A simple scaling is to divide Σ_i at each i by its maximum value in $0 \leqslant x \leqslant 1$, so that

$$S_i(x) = \bar{\lambda}_i \Sigma_i(x) \tag{4.144}$$

where

$$(\bar{\lambda}_i)^{-1} = \max_{0 \leqslant x \leqslant 1} \Sigma_i(x). \tag{4.145}$$

According to the computations carried out by Berryman, this normalisation, though trivial as far as computer time is concerned, is crucial for avoiding divergence and strong dependence on the choice of initial iterate. Indeed, Luning and Perry (1981) have rigorously proved that this iterative process converges for a suitable choice of the initial iterate. Berryman has chosen the initial iterate $S_0(x) = \phi(x) \geqslant 0$ such that ϕ is a continuous integrable function not identically zero.

Now, the Green's function for eq. (3.143) (Birkhoff & Rota, 1978, p. 46) is

$$G(x, \xi) = \begin{Bmatrix} x(1-\xi), & x \leqslant \xi, \\ \xi(1-x), & x \geqslant \xi. \end{Bmatrix} \tag{4.146}$$

The explicit solution of eq. (4.143) then is

$$S_i(x) = \lambda_i \int_0^1 G(x, \xi) F(\xi) S_{i-1}^{q-1}(\xi) \mathrm{d}\xi. \tag{4.147}$$

The specific initial choice $S_0 = 4x(1-x)$, satisfying the zero boundary conditions at $x = 0$ and $x = 1$, turns out to be quite convenient, and all the computations by Berryman (1977) and Drake and Berryman (1977), for different values of q, were found to converge rapidly.

Eq. (4.147) shows that the geometrical factor $F(x)$ can make $S(x)$ asymmetric if $F(x)$ itself is. It can change the magnitude of the eigenvalue λ and hence the time decay rate of the solution, T^q, since $q = 2\lambda/\varrho^2$. More importantly, it can affect the stability of the separable solution if the decay rates of the perturbations to this solution get altered substantially more than the original decay rate.

Before studying the stability of the separable solution and its asymptotic behaviour, we summarise the results for the numerical solutions of the original partial differential eq. (4.124) with $F(x) = 1, \delta = -\frac{1}{2}$, and initial conditions

$$m(x, 0) = \sum_{l=0}^{3} a_l \sin(l+1)x. \tag{4.148}$$

Four different sets of values for the Fourier coefficients (a_0, a_1, a_2, a_3) were chosen: (i) $(1, 0.4, 0, 0)$, (ii) $(1, 0, 0.3, 0)$, (iii) $(1, 0, -0.3, 0)$ and (iv) $(1, 0, 0, 0.225)$. Eq. (4.124) was integrated using a linear three-level difference scheme developed by Lees (1966). By times $t \lesssim 0.1$, the particle distribution for all four cases decayed and began to evolve according to the separable solution. In all cases, the particles escape ($n = 0$) before $t = 0.2$. Thus, several initial density distributions, finitely different from the separable solution, converge to the latter in a short time, confirming the experimental observations that particles, initially injected into the containment device with some arbitrary spatial distribution, after a finite time appear essentially in the normal mode, that is, the separable solution. Here, then, is an example of a product (or separable) solution serving as an intermediate asymptotic – akin to but different from the similarity solutions discussed in earlier sections.

Stability and asymptotic behaviour of the separable solution

First we consider the case with $F(x) = 1$, namely eq. (4.129), and assume the perturbed solution to have the product form

$$m(x,t) = S(x)T(t) + u(x)v(t) \tag{4.149}$$

where $T(t)$ and $S(x)$ are given by eqs. (4.132) and (4.133) and the term $u(x)v(t)$ is small. Substituting (4.149) into eq. (4.129) and linearising, we can write the resulting equation in a separable form

$$(q-1)T^{q-2}[\ln(vT^{q-2})]_t = u_{xx}S^{2-q}/u = -\kappa, \tag{4.150}$$

κ being the separation constant. The equation for v,

$$(q-1)T^{q-2}[\ln(vT^{q-2})]_t + \kappa = 0, \tag{4.151}$$

has a solution of the form

$$v = cT^p, \tag{4.152}$$

where c is a constant, if

$$p + q - 2 = \frac{\kappa}{\lambda}. \tag{4.153}$$

Since T decreases as t increases (see eq. (4.132)), v decreases if $p > 0$ and decays faster than the separable solution if $p > 1$. Therefore, to ensure stability, we require that $p > 1$. To find p, we have to find κ_i from the eigenvalue problem

$$u_i''(x) + \kappa_i S^{q-2}(x)u_i(x) = 0, \qquad u_i(0) = 0, \quad u_i(1) = 0. \tag{4.154}$$

For $\kappa_0 = \lambda$, the solution u_0 is simply S. This solution is positive everywhere. The general eigenvalue problem (4.154) (Coddington & Levinson, 1955) has an infinite number of eigenvalues $\kappa_0, \kappa_1, \ldots$ with $\kappa_0 \leqslant \kappa_1 \leqslant \kappa_2 \ldots$ and $\kappa_i \to \infty$ as $i \to \infty$. The eigenfunction u_i corresponding to κ_i has exactly i zeros in $(0, 1)$. We can now identify κ_0 and u_0 with the lowest eigenvalue and the corresponding eigenfunction, respectively. The higher eigenfunctions $u_1, u_2, \ldots$ have exactly one, two, ... zeros, respectively, in $(0, 1)$. The perturbation u_0, we have seen, corresponds to a change in the initial amplitude of the separable solution. For this case ($\kappa_0 = \lambda$), eq. (4.153) gives $p = 3 - q$. Therefore, according to eq. (4.125),

$$p \geqslant \begin{Bmatrix} 1, & \delta \geqslant 0, \\ 0, & \delta \geqslant -\frac{1}{2}. \end{Bmatrix} \tag{4.155}$$

This implies stability of the separable solution for $\delta \geqslant 0$ with respect to the lowest mode. For $0 \geqslant \delta \geqslant -1$, further examination is needed.

Now, turning to κ_1 and u_1, we seek a solution of (4.154) which vanishes only once in $(0, 1)$ and also vanishes at the boundaries. A trial solution

$$u_1 = S(x)S'(x), \tag{4.156}$$

which meets both the requirements (since $S(x)$ vanishes at the boundaries and $S'(x) = 0$ at $x = \frac{1}{2}$), is easily seen, on use of eq. (4.136), to satisfy eq. (4.154) with eigenvalue

$$\kappa_1 = q(\lambda + \varrho^2) = (2 + q)\lambda. \tag{4.157}$$

Substituting κ_1/λ from eq. (4.157) into eq. (4.153) we find that

$$p_1 = 4 \tag{4.158}$$

for all q. Thus, the lowest nontrivial perturbation decays much faster than the separable solution. All other eigenvalues decay still faster, since $\kappa_i/\lambda > \kappa_1/\lambda$ for $i > 1$. The stability of the separable solution has therefore been established in the geometry-free case, $F(x) = 1$.

Berryman has also proved stability of the separable solution in the presence of a geometrical factor $F(x)$ for $\delta \geqslant -\frac{1}{2}$, provided $F(x)$ is symmetric about $x = \frac{1}{2}$ and $F'(x) \geqslant 0$ for $0 \leqslant x \leqslant \frac{1}{2}$. It is conjectured that the separable solution is stable for $\delta > -1$ for arbitrary functions $F(x)$ which satisfy the conditions (4.127) and (4.128) and for which $F'(x) \geqslant 0$ for $x \leqslant x_s$. This conjecture is supported by numerical evidence.

Now we derive bounds on the asymptotic amplitude A_0 for large t, for slow diffusion, $1 < q < 2$, corresponding to $\delta > 0$, and the extinction time t^* of the solution for fast diffusion, $q > 2$, for which the density vanishes everywhere in a finite time (Berryman & Holland, 1978). It is useful to introduce the following integrals:

$$a_0(t) = c^{-1} \int m(x,t) F(x) S^{q-1}(x) \mathrm{d}x, \tag{4.159}$$

$$\beta(t) = c^{-1} \int m^{q-1}(x,t) F(x) S(x) \mathrm{d}x, \tag{4.160}$$

$$Q(t) = c^{-1} \int m^{q}(x,t) F(x) \mathrm{d}x, \tag{4.161}$$

$$R(t) = c^{-1} \int m_x^2(x,t) \, \mathrm{d}x, \tag{4.162}$$

where

$$c = \int F(x) S^{q}(x) \mathrm{d}x \tag{4.163}$$

and all integrals are taken over the interval $0 \leqslant x \leqslant 1$. The function $F(x)$ (see (4.127) and (4.128)) is positive and integrable.

If we differentiate eq. (4.160) with respect to t, use eq. (4.124) to replace $F(x)\,(m^{q-1})_t$ by m_{xx} and integrate by parts employing the null boundary conditions on $m(x,t)$ and $S(x)$ at $x = 0, 1$, we have

$$\frac{\mathrm{d}}{\mathrm{d}t}\beta(t) = c^{-1} \int m_{xx} S \, \mathrm{d}x = c^{-1} \int m S_{xx} \, \mathrm{d}x$$

$$= -\lambda c^{-1} \int m(x,t) F(x) S^{q-1}(x) \mathrm{d}x = -\lambda a_0(t). \tag{4.164}$$

In the penultimate step, we have made use of eq. (4.142). Eq. (4.164) provides one relation between $\beta(t)$ and $a_0(t)$. A second relation is obtained by Hölder's inequality applied to the integral in eq. (4.159) for (positive) $f(x)$ and $g(x)$, namely

$$\int f(x) g(x) \mathrm{d}x \leqslant \left\{ \int [f(x)]^P \mathrm{d}x \right\}^{1/P} \left\{ \int [g(x)]^Q \mathrm{d}x \right\}^{1/Q}, \tag{4.165}$$

with $1/P + 1/Q = 1$. For the integral in eq. (4.159), we have

$$f(x) = mF^{1/(q-1)} S^{1/(q-1)}, \quad g(x) = F^{(q-2)/(q-1)} S^{q(q-2)/(q-1)},$$

$$P = q-1, \quad Q = \frac{q-1}{q-2}, \quad \frac{1}{P} + \frac{1}{Q} = 1,$$

so that, using the definition (4.163) for c,

$$a_0(t) \leqslant [\beta(t)]^{1/(q-1)}. \tag{4.166}$$

Eliminating $a_0(t)$ from eqs. (4.164) and (4.166), we obtain a differential inequality for $\beta(t)$:

$$-\frac{\mathrm{d}}{\mathrm{d}t}\beta(t) \leqslant \lambda[\beta(t)]^{1/(q-1)}. \tag{4.167}$$

Upon integration of eq. (4.167) and changing the sign etc., we have

$$\beta^r(t) \geqslant \beta^r(0) - \lambda r t, \tag{4.168}$$

where

$$r = \frac{q-2}{q-1}. \tag{4.169}$$

For fast diffusion, $-1 < \delta < 0, q > 2$, implying $r > 0$ (see eq. (4.125)), if the pseudo-density m (and hence the density n) vanishes after a finite time t^* (see eq. (4.132)), then $\beta(t^*)$ vanishes too. Therefore, the extinction time t^*, according to (4.168), satisfies the inequality

$$t^* \geqslant \frac{\beta^r(0)}{\lambda r}. \tag{4.170}$$

Thus, (4.170) provides a rigorous lower bound for t^*.

For slow diffusion, $1 < q < 2, r < 0$; therefore, it follows from (4.168) that

$$\beta^{|r|}(t) \leqslant [\beta^r(0) + \lambda |r| t]^{-1}. \tag{4.171}$$

In the asymptotic time limit, $m(x,t) \to S(x)T(t)$ (see Berryman and Holland (1978)), eq. (4.132) gives

$$T^{q-2} = A_0^{q-2} + \lambda |r| t. \tag{4.172}$$

Substituting $S(x)\ T(t)$ for $m(x,t)$ in eq. (4.160) and employing the definition (4.163) for c, we evaluate $\beta^{|r|}(t)$; hence (4.171) and (4.172) give

$$(A_0^{q-2} + \lambda |r| t)^{-1} \leqslant [\beta^r(0) + \lambda |r| t]^{-1},$$

that is,

$$A_0 \leqslant [\beta(0)]^{1/(q-1)}. \tag{4.173}$$

The inequality (4.173) provides an upper bound for the asymptotic amplitude A_0.

We now derive an upper bound for the extinction time t^* for fast diffusion, $q > 2$. By the definition of $Q(t)$ and $R(t)$, eqs. (4.161)–(4.162), and use of eq. (4.124) for m, we have

$$\frac{\mathrm{d}Q}{\mathrm{d}t} = -\frac{qR(t)}{q-1} \leqslant 0, \tag{4.174}$$

$$\frac{dR}{dt} = -\frac{2c^{-1}}{q-1}\int \frac{m_{xx}^2}{m^{q-2}F(x)}\,dx \leqslant 0, \tag{4.175}$$

and

$$R(t) = -c^{-1}\int m_{xx}m\,dx = -c^{-1}\int\left(\frac{m_{xx}}{m^{q/2-1}F^{1/2}}\right)(m^{q/2}F^{1/2})\,dx. \tag{4.176}$$

In (4.175) and (4.176), we have performed integration by parts once. We now apply Schwarz's inequality to eq. (4.176) to obtain

$$R^2(t) \leqslant c^{-2}\int m^q F\,dx \int \frac{m_{xx}^2\,dx}{m^{q-2}F} = Q(t)c^{-1}\int \frac{m_{xx}^2}{m^{q-2}}\frac{dx}{F}. \tag{4.177}$$

Eliminating the integral from (4.175) and (4.177) and combining the result with (4.174), we have the inequality

$$\frac{d}{dt}[R(t)Q^{-2/q}] \leqslant 0. \tag{4.178}$$

It is easily checked that the equality sign in (4.178) holds only when we use the separable solution $m = S(x)\,T(t)$. Again substituting this limiting expression for m into eqs. (4.161) and (4.162), we have

$$\begin{aligned} R(t)Q^{-2/q} &\geqslant c^{-1}\int S_x^2 T^2(t)\,dx \cdot c^{2/q}\left(\int S^q F\,dx\right)^{-2/q} T^{-2} \\ &= c^{-1}\lambda c c^{2/q} c^{-2/q} = \lambda, \end{aligned} \tag{4.179}$$

since

$$\int S_x^2\,dx = -\int S_{xx}S\,dx = \lambda\int FS^q\,dx = \lambda c. \tag{4.180}$$

The inequality (4.178), by an integration and use of (4.179), gives

$$\lambda Q^{2/q} \leqslant R(t). \tag{4.181}$$

Eliminating $R(t)$ from (4.174) and (4.181) and integrating, we have

$$-\frac{1}{r}[Q^{(q-2)/q}(t) - Q^{(q-2)/q}(0)] \geqslant \lambda t. \tag{4.182}$$

Since for fast diffusion, $q > 2, r > 0$, and at the extinction time $t = t^*$, $Q(t^*) = 0$, we have, from (4.182),

$$t^* \leqslant [Q(0)]^{(q-2)/q}/\lambda r. \tag{4.183}$$

This provides an upper bound for t^*. The bounds (4.170), (4.173) and (4.183) are the best possible since the equality sign holds when $m(x,t)$ is the separable solution.

Berryman and Holland (1978) have compared these bounds with their perturbation solution, using a set of initial data. The perturbation solution was found to lie between the upper and the lower bound.

In another paper, Berryman and Holland (1982) have considered the case $\delta = -1$, $n_t = (\ln n)_{xx}$, and have proved the asymptotic form of solution, namely $\ln[n(x,t)/n_0] \to A\,\mathrm{e}^{-\pi^2 t/n_0}\sqrt{2}\sin \pi x$ as $t \to \infty$, where n_0 is the value of n at the boundaries of the interval $[0, 1]$ and A is the asymptotic amplitude.

4.6 The non-planar Burgers equation

We shall now discuss the similarity solution of the generalised Burgers equation

$$u_x - uu_t = g(x)u_{tt} \tag{4.184}$$

whose special planar case we introduced in sec. 3.5. Here, t is the retarded time, $t = T - R/c_0$, $x = [(\gamma+1)/2c_0^2]\int a^{-1/2}(R)\,\mathrm{d}R$ and $v = a^{-1/2}u$ is the particle velocity. The function $g(x)$ is determined by the duct area $a(R)$ (a function of the actual distance R); $g(x) = \delta a^{1/2}(R)/(\gamma+1)c_0$. The constants δ, c_0 and γ stand for the coefficient of diffusivity of sound, the undisturbed sound speed and the ratio of specific heats, respectively. The function $g(x)$ represents the effect of geometrical expansion, which for spherical, cylindrical and plane symmetries is e^x, x and 1, respectively. As we explained in sec. 3.5, the role of space and time coordinates has been interchanged in writing eq. (4.184) to facilitate the posing of a boundary value problem. Thus x is, in fact, a time-like variable and the boundary condition for eq. (4.184) is $u(0,t) = u_0(t)$, simulating a piston motion. Before we discuss the asymptotic behaviour of eq. (4.184), we analyse its similarity solution for the cylindrical case $g(x) = \beta x$ where β is a constant. This solution is an intermediate asymptotic (Rudenko & Soluyan (1977); see also sec. 5.8).

We easily check that the solution to the equation

$$u_x - uu_t = \beta x u_{tt} \tag{4.185}$$

has a self-similar form

$$u = \Omega(t/x) = \Omega(\xi). \tag{4.186}$$

Substituting (4.186) into eq. (4.185) we have

$$-(\Omega + \xi)\Omega' = \beta\Omega''. \tag{4.187}$$

We simplify this equation by introducing

$$\Omega + \xi = \beta^{1/2} H, \quad \xi = \beta^{1/2}\zeta \tag{4.188}$$

so that it becomes

$$\left(1 - \frac{\mathrm{d}H}{\mathrm{d}\zeta}\right) H = \frac{\mathrm{d}^2 H}{\mathrm{d}\zeta^2}. \tag{4.189}$$

This equation is autonomous and easily admits a first integral. However, this integral is not appropriate to the present problem wherein we want the solution to tend to constant values, say, $C \pm D$ as $\zeta \to \pm\infty$; these boundary conditions result from a piston motion at $x = \text{const}$ with constant end conditions at $t \to \pm\infty$. We transform eq. (4.189) by changing both the dependent and independent variables,

$$H = 2F(\eta(\zeta)), \tag{4.190}$$

so that it becomes

$$F''\left(\frac{\mathrm{d}\eta}{\mathrm{d}\zeta}\right)^2 + F'\frac{\mathrm{d}^2\eta}{\mathrm{d}\zeta^2} = F\left(1 - 2F'\frac{\mathrm{d}\eta}{\mathrm{d}\zeta}\right), \tag{4.191}$$

the prime denoting differentiation with respect to η. Now $\mathrm{d}\eta/\mathrm{d}\zeta$ is chosen such that differentiated terms have the same degree of nonlinearity; this leads to $\mathrm{d}\eta/\mathrm{d}\zeta = F$ and eq. (4.191) becomes

$$FF'' + F'(F' + 2F) = 1. \tag{4.192}$$

Setting $F^2 = K$, we have a linear equation

$$\frac{K''}{2} + K' = 1 \tag{4.193}$$

which has the solution

$$K = \eta + c\,\mathrm{e}^{-2\eta} = \eta - \eta_0\,\mathrm{e}^{-2(\eta - \eta_0)}, \text{ say}, \tag{4.194}$$

so that

$$F = \pm(\eta - \eta_0\,\mathrm{e}^{-2(\eta - \eta_0)})^{1/2}. \tag{4.195}$$

Here c and η_0 are constants. The solution for u is written in a parametric form as

$$\left.\begin{aligned} u &= \Omega(t/x) = -\xi + 2\beta^{1/2} F(\eta), \\ \xi &= \gamma + \beta^{1/2}\int_{\eta_0}^{\eta} \frac{\mathrm{d}\eta}{F(\eta)}, \end{aligned}\right\} \eta \geqslant \eta_0, \tag{4.196}$$

where η is the parameter. The second of eqs. (4.196) follows from integrating $\mathrm{d}\eta/\mathrm{d}\zeta = F$ and assuming $\xi = \gamma$ when $\eta = \eta_0$. The solution (4.196) has two constant parameters, γ and η_0. While γ can assume any real value, the parameter η_0 can vary only between $-\frac{1}{2}$ and $+\infty$. The inequalities $\eta \geqslant \eta_0 > -\frac{1}{2}$ follow easily from the consideration of eqs. (4.195). The choice of sign in eq. (4.195) gives two sections of the curve for $\Omega(\xi)$: the plus sign corresponds to the curve with $\xi \geqslant \gamma$ while the minus sign defines it for $\xi \leqslant \gamma$. The two parts join smoothly at $\xi = \gamma$, corresponding to $\eta = \eta_0$. This can be seen by expanding F^2 about $\eta = \eta_0$: to a first approximation, $F^2 \approx (1 + 2\eta_0)\cdot(\eta - \eta_0)$. The parameter γ merely reflects the Galilean invariance of eq. (4.185), $u \to u + \lambda, t \to t + \lambda x$; the parameter η_0 plays a crucial role. In fact, it is a measure of the strength of the wave: as $\xi \to \pm\infty$ (corresponding to $\eta \to \infty, -\frac{1}{2}$, respectively), $\Omega \to -\gamma \pm \beta^{1/2}\Delta(\eta_0)$. The strength of the wave, $2\beta^{1/2}\Delta(\eta_0)$, depends on η_0 and increases from $-\infty$ to $+\infty$ as η_0 increases from $-\frac{1}{2}$ to $+\infty$, so that the wave can have any height $2\beta^{1/2}\Delta(\eta_0)$, positive or negative. Since $\mathrm{d}\Omega/\mathrm{d}\xi = 2\eta_0\,\mathrm{e}^{-2(\eta-\eta_0)}$, the profile $\Omega(\xi)$ is monotonically increasing or decreasing, depending on whether η_0 is positive or negative.

It is easy to find particular solutions of eq. (4.187) relevant to other special circumstances. Thus, when the solution is essentially linear, corresponding to $\eta_0 \to 0$, we may ignore the $\Omega\Omega'$ term in eq. (4.187) so that

$$-\xi\Omega' = \beta\Omega''. \tag{4.197}$$

This equation can be integrated to give

$$\Omega = c_0\left(\frac{2\pi}{\beta}\right)^{1/2} \operatorname{erf}\frac{\xi - \gamma}{(2\beta)^{1/2}} + c_1. \tag{4.198}$$

This can be matched to eq. (4.196) at $\eta = \eta_0$ in the limit $\eta_0 \to 0$ by choosing c_0 and c_1 suitably; we have then

$$\Omega = (2\pi\beta)^{1/2}\eta_0 \operatorname{erf}[(\xi - \gamma)/(2\beta)^{1/2}] - \gamma. \tag{4.199}$$

Similarly, if we seek a Taylor-shock solution, we balance nonlinear and viscous terms in eq. (4.187):

$$-\Omega\Omega' = \beta\Omega''. \tag{4.200}$$

This equation when integrated and matched to eq. (4.196) at $\eta = \eta_0$ in the limit $\eta_0 \to \infty$ gives

$$\Omega = 2(\beta\eta_0)^{1/2} \tanh[(\eta_0/\beta)^{1/2}(\xi - \gamma)] - \gamma. \tag{4.201}$$

The role of η_0 as a measure of the wave strength has now been clarified. Finally, $\Omega = -\xi$ is obviously a solution of eq. (4.187). This solution, holding in a finite interval, can be joined to constant solutions of eq. (4.187) at the two ends of the interval to give an expansion front.

We refer here to a related equation

$$\varepsilon x y'' + (g(x) - y)y' = 0, \quad x \in (0, R)$$
$$y(0) = 0, \quad y(R) = k,$$

which describes stationary solutions of a corresponding evolution equation (see sec. 4.9). The latter arises in the context of pre-breakdown discharge in an ionised gas between two electrodes. Diekmann *et al.* (1980) have studied this equation in detail with particular reference to the physical nature of $g(x)$ and the limiting behaviour of the solution as the diffusivity coefficient $\varepsilon \to 0$ and the domain $R \to \infty$. Matched perturbation solutions were constructed and their asymptotic nature was analysed.

4.7 Stability of the self-similar solution of the cylindrically symmetric Burgers equation – another intermediate asymptotic

Now we study the solutions of initial value problems

$$u_x - uu_t = g(x)u_{xx}, \quad 0 \leqslant x < \infty, \quad g(x)/x \to \beta \neq 0 \ \text{ as } x \to \infty, \tag{4.202}$$

$$u(0, t) = u_0(t), \tag{4.203}$$

which tend asymptotically to the similarity solution $u = \Omega(t/x)$ of eq. (4.185) as x tends to infinity (Scott, 1981b). We have already found the solution $\Omega(t/x)$ in sec. 4.6. We now assume that $u_0(t)$ is continuous and bounded and is such that a unique bounded solution to eqs. (4.202)–(4.203) exists, for which u, u_t, u_x, and u_{tt} are continuous in $x > 0$. We shall also assume that if $u_0(t)$ satisfies the conditions

$$u_0(t) \to C \pm D \quad \text{as } t \to \pm\infty, \tag{4.204}$$

and

$$\frac{du_0}{dt} \text{ is continuous and tends to 0 as } t \to \pm\infty,$$

then $u(x, t) \to C \pm D$ and $\partial u/\partial t \to 0$, as $t \to \pm\infty$, uniformly in $0 \leqslant x \leqslant x_1 < \infty$.

We shall prove that the solution $u(x, t)$ of eqs. (4.202)–(4.203) under the above conditions satisfies the asymptotic relation

$$|\Omega(t/x) - u(x, t)| \to 0 \tag{4.205}$$

as $x \to \infty$, uniformly in t. Here, the asymptotic behaviour of the initial profile $u_0(t)$ (and hence of $u(x,t)$) as $t \to \infty$ is made to coincide with that of $\Omega(t/x)$ (see the discussion following eq. (4.196)). Accordingly, we determine that $\gamma = -C$ and $\beta^{1/2}\Delta(\eta_0) = D$. The proof of (4.205) is carried out in three steps. First, (4.205) is proved with the initial condition $u_0(t) = \Omega(t)$, the functional form of the similarity solution itself (cf. sec. 4.6). Secondly, $u_0(t)$ is extended to become $\Omega(t) + A\,\mathrm{sech}\,t$, where A is constant. Finally, $u_0(t)$ is chosen to be any continuous function satisfying (4.204).

In contrast to the results for the nonlinear heat equation in sec. 4.4 and in view of the more complicated nature of eq. (4.202), no estimates for the rates of convergence are obtained. The proof for convergence will involve some estimates for the fundamental solution of linear parabolic equations with variable coefficients, for which we shall refer to Friedman (1964). The notation F(*a*. *b*. *c*) will refer to theorem or lemma number *c* of § *b* in chapter *a* of Friedman (1964).

Lemma 1. The asymptotic result (4.205) holds when $u_0(t) = \Omega(t)$.

Proof. Let $v = \Omega(t/[x+1])$ be the (similarity) solution of

$$v_t - vv_x = \beta(x+1)v_{tt}. \tag{4.206}$$

u satisfies the equation

$$u_t - uu_x = g(x)u_{tt}. \tag{4.207}$$

Subtracting eq. (4.206) from eq. (4.207) and writing $w = u - v$, we have

$$\frac{\partial w}{\partial x} - \frac{1}{2}\frac{\partial}{\partial t}[w(u+v)] = g(x)\frac{\partial^2 w}{\partial t^2} + [g(x) - \beta(x+1)]\frac{\partial^2 v}{\partial t^2}, \tag{4.208}$$

where

$$w(0,t) = u(0,t) - v(0,t) = 0. \tag{4.209}$$

By introducing the independent variable

$$z = \int_0^x g(\xi)\,\mathrm{d}\xi \tag{4.210}$$

and writing $g(x) = G(z)$, eq. (4.208) becomes

$$\frac{\partial w}{\partial z} - \frac{\partial}{\partial t}\left[\frac{w(u+v)}{2G(z)}\right] - \frac{\partial^2 w}{\partial t^2} = \frac{G(z) - \beta(x+1)}{G(z)}\frac{\partial^2 v}{\partial t^2}, \tag{4.211}$$

where $v = \Omega(t/[x+1])$ and $(x+1)$ are known functions of z and t so that eq. (4.211) is a linear inhomogeneous equation in w with the source term

$$\frac{G-\beta(x+1)}{G}\frac{\partial^2 v}{\partial t^2}=\frac{\partial H}{\partial t}, \tag{4.212}$$

where

$$H(z,t)=\frac{G-\beta(x+1)}{G}\frac{\partial v}{\partial t}. \tag{4.213}$$

Here, the coefficient of $\partial v/\partial t$ is a function of z alone. If we write $w=\partial\psi/\partial t$ in eq. (4.211) and integrate it with respect to t, ignoring the 'constant' of integration, we have

$$\frac{\partial\psi}{\partial z}-\frac{(u+v)}{2G}\frac{\partial\psi}{\partial t}-\frac{\partial^2\psi}{\partial t^2}=H(z,t). \tag{4.214}$$

We write eq. (4.214) in the form

$$L\phi\equiv\frac{\partial\phi}{\partial z}-\frac{(u+v)}{2G}\frac{\partial\phi}{\partial t}-\frac{\partial^2\phi}{\partial t^2}=H(z,t). \tag{4.215}$$

If $\Gamma(z,t;\zeta,\tau)$ is the fundamental solution of $L\phi=0$, then it easily follows that the solution ψ of eq. (4.214) is given by

$$\psi(z,t)=\int_0^z\int_{-\infty}^{\infty}\Gamma(z,t;\zeta,\tau)H(\zeta,\tau)\mathrm{d}\tau\,\mathrm{d}\zeta \tag{4.216}$$

with $\psi(0,t)=0$ (see F (1.7.12)). The partial derivative of the fundamental solution satisfies the inequality

$$\frac{\partial\Gamma}{\partial t}\leqslant K_1\frac{\mathrm{e}^{-K_2(t-\tau)^2/(z-\zeta)}}{z-\zeta}, \tag{4.217}$$

where K_1 and K_2 are constants, independent of z,t,ζ and τ (see the derivation of F (1.6.12)). Since $\partial v/\partial t$ is bounded and $g(x)\to\beta x$ as $x\to\infty$ (that is, as $z\to\infty$), it follows from eqs. (4.213) and (4.210) that

$$H(z,t)=o(x^{-1})=o(z^{-1/2}) \tag{4.218}$$

as $z\to\infty$, uniformly in t.

If we define

$$I(z)=\int_0^z\max_{-\infty<t<\infty}|H|\,\mathrm{d}z, \tag{4.219}$$

then by theorem F (2.4.9), wherein

$$L=\left(\frac{\partial^2}{\partial t^2}+\frac{u+v}{2G}\frac{\partial}{\partial t}-\frac{\partial}{\partial z}\right) \tag{4.220}$$

and the relevant functions are

$$I(z) \pm \psi(z,t), \tag{4.221}$$

we obtain

$$|\psi| \leqslant I(z) = o(z^{1/2}) \tag{4.222}$$

as $z \to \infty$, uniformly in t.

Eq. (4.216) may be rewritten as

$$\psi(z,t) = \int_{-\infty}^{\infty} \Gamma\left(z,t;\frac{z}{2},\tau\right)\psi\left(\frac{z}{2},\tau\right)\mathrm{d}\tau + \int_{z/2}^{z}\int_{-\infty}^{\infty} \Gamma(z,t;\zeta,\tau)H(\zeta,\tau)\mathrm{d}\tau\,\mathrm{d}\zeta, \tag{4.223}$$

by making use of (4.29) of F (1.4.8).

Differentiating eq. (4.223) with respect to t, using the estimates (4.217), (4.218) and (4.222) for $\partial\Gamma/\partial t$, H and ψ, respectively, we have

$$|w| = \left|\frac{\partial\psi}{\partial t}\right| \leqslant \int_{-\infty}^{\infty}\left|\frac{\partial\Gamma}{\partial t}(z,t;z/2,\tau)\right|\left|\psi\left(\frac{z}{2},\tau\right)\right|\mathrm{d}\tau + \int_{z/2}^{z}\int_{-\infty}^{\infty}\left|\frac{\partial\Gamma}{\partial t}(z,t;\zeta,\tau)\right||H(\zeta,\tau)|\,\mathrm{d}\tau\,\mathrm{d}\zeta = o(1), \tag{4.224}$$

the two terms on the right side of (4.224) being of order $|\psi|/z^{1/2}$ and $1/z^{\varepsilon}, \varepsilon > 0$, respectively, and hence small as $z \to \infty$, uniformly in t.

Since $w = u(x,t) - \Omega(t/(x+1))$, we have

$$|u(x,t) - \Omega(t/x)| \leqslant |u(x,t) - \Omega(t/(x+1))| + |\Omega(t/[x+1]) - \Omega(t/x)| \to 0. \tag{4.225}$$

as $x \to \infty$, uniformly in t.

We now proceed to the second stage of the proof, in which it is shown that all the solutions of eq. (4.202) corresponding to the initial conditions

$$u_0(t) = \Omega(t) + A\,\mathrm{sech}\,t, \quad A \text{ a constant}, \tag{4.226}$$

converge to that with $\Omega(t)$ as the initial condition; we denote the latter solution by $U(x,t)$. We first prove the following lemmas.

Lemma 2. If $u_0(t)$ has the form (4.226), then $|\partial u/\partial t| < K/x$ for some constant K, for all t and $x \geqslant 0$.

Proof. We write the basic equation (4.202) as

$$\frac{\partial \bar{w}}{\partial z} - \frac{u}{G}\frac{\partial \bar{w}}{\partial t} = \frac{\partial^2 \bar{w}}{\partial t^2} \tag{4.227}$$

so that $\bar{w}$ may be thought of as a solution of this linear equation (z here is as in eq. (4.210)). Let the fundamental solution of this equation be again denoted by $\Gamma(z, t; \zeta, \tau)$, for which the inequality (4.217) holds. The homogeneous eq. (4.227) has the 'solution' (cf. eq. (4.223))

$$u(z, t) = \int_{-\infty}^{\infty} u\left(\frac{z}{2}, \tau\right)\Gamma\left(z, t; \frac{z}{2}, \tau\right)\mathrm{d}\tau \tag{4.228}$$

so that

$$\left|\frac{\partial u}{\partial t}\right| \leqslant \int_{-\infty}^{\infty}\left|u\left(\frac{z}{2}, \tau\right)\right|\left|\frac{\partial \Gamma(z, t; z/2, \tau)}{\partial t}\right|\mathrm{d}\tau \tag{4.229}$$

$$\leqslant \text{constant} \times \int_{-\infty}^{\infty}\left|\frac{\partial \Gamma}{\partial t}(z, t; z/2, \tau)\right|\mathrm{d}\tau$$

$$\leqslant \frac{K}{x}, \tag{4.230}$$

using (4.217) and the boundedness of $u(z/2, \tau)$, and integrating. Hence the lemma is established.

Lemma 3. If $u_0(t)$ is of the form (4.226), then

$$|U(x, t) - u(x, t)| \to 0 \tag{4.231}$$

as $x \to \infty$, uniformly in t. We recall that we have here denoted the solution of eq. (4.202) with $u_0(t) = \Omega(t)$ by $U(x, t)$.

Proof. We again write $w = u - U$ so that (cf. eq. (4.211))

$$\frac{\partial w}{\partial x} - \frac{\partial}{\partial t}[\tfrac{1}{2}w(u + U)] = g(x)\frac{\partial^2 w}{\partial t^2}. \tag{4.232}$$

First we assume that $A \geqslant 0$, so that $w(0, t) = A \operatorname{sech} t \geqslant 0$, then, by theorem F (2.4.9), $w \geqslant 0$ for all t and for $x \geqslant 0$. Further, by Lemma 2,

$$\left|\frac{\partial w}{\partial t}\right| \leqslant \frac{K}{x} \tag{4.233}$$

for some constant K.

Integrating eq. (4.232) with respect to t from t_1 to t_2, we get

$$\frac{d}{dx}\int_{t_1}^{t_2} w\,dt = \left[g(x)\frac{\partial w}{\partial t} + \frac{1}{2}(u+U)w \right]_{t_1}^{t_2}. \tag{4.234}$$

Integrating (4.234) with respect to x and remembering that $w(0,t) = A$ sech t, we have

$$\int_{t_1}^{t_2} w\,dt = \int_0^x \left[g(x)\frac{\partial w}{\partial t} + \frac{1}{2}(u+U)w \right]_{t_1}^{t_2} dx + 2A[\arctan e^t]_{t_1}^{t_2}, \tag{4.235}$$

the last term on the right being the value of the integral on the left at $x = 0$. In the limits $t_1 \to -\infty$ and $t_2 \to \infty$, w and $\partial w/\partial t$ both tend to zero by basic assumptions about the solutions of eq. (4.202) and the initial conditions. Therefore, we find that

$$\int_{-\infty}^{\infty} w\,dt = A\pi \tag{4.236}$$

for all x.

For a fixed x, let the maximum value of w be w_{max} at $t = t_{max}$. Considering the two parts of the inequality (4.233) separately and integrating with respect to t, we have

$$w \geqslant w_{max} - K|t - t_{max}|/x. \tag{4.237}$$

Integrating (4.237) in the interval $|t - t_{max}| \leqslant xw_{max}/K$, we have

$$\int_{|t-t_{max}| \leqslant xw_{max}/K} w\,dt \geqslant \int_{-xw_{max}/K}^{xw_{max}/K} \left(w_{max} - \frac{K|z|}{x} \right) dz = \frac{xw_{max}^2}{K}. \tag{4.238}$$

We have shown earlier that $w(x,t) \geqslant 0$ (see opposite eq. (4.232)), and so combining (4.236) and (4.238) we have

$$A\pi \geqslant xw_{max}^2/K \tag{4.239}$$

or

$$w_{max} \leqslant (A\pi K/x)^{1/2} \to 0 \tag{4.240}$$

as $x \to \infty$. Thus, we have

$$|w| = |u - U| \to 0 \tag{4.241}$$

as $x \to \infty$, uniformly in t.

The proof for $A \leqslant 0$ is similar. Hence the lemma follows. Finally, we prove the main theorem of this section.

Theorem 1. The result (4.205) remains valid for any continuous function $u_0(t)$ satisfying (4.204).

Proof. The analysis for this part of the theorem is similar to that for the nonlinear heat equation (sec. 4.4). For any $\varepsilon > 0$, we can choose $A > 0$ such that

$$|u_0(t) - \Omega(t)| \leqslant \varepsilon + A \operatorname{sech} t. \tag{4.242}$$

We construct two solutions $u_{\pm}(x,t)$ of eq. (4.202) with initial conditions

$$u_{\pm}(0,t) = \Omega(t) \pm (\varepsilon + A \operatorname{sech} t). \tag{4.243}$$

Then the difference functions

$$w_{\pm}(x,t) = \pm(u_{\pm} - u) \tag{4.244}$$

satisfy the equations

$$\frac{\partial w_{\pm}}{\partial x} - u\frac{\partial w_{\pm}}{\partial t} - \frac{\partial u_{\pm}}{\partial t} w_{\pm} = g(x)\frac{\partial^2 w_{\pm}}{\partial t^2}, \tag{4.245}$$

and $w_{\pm}(0,t) \geqslant 0$ in view of (4.242)–(4.244). Hence by theorem F (2.4.9) we have $w_{\pm}(x,t) \geqslant 0$ for all t, and $x \geqslant 0$. It follows, therefore, that

$$u_- \leqslant u \leqslant u_+. \tag{4.246}$$

Because of the Galilean invariance of eq. (4.202) (see sec. 4.6), the functions

$$v_{\pm}(x,t) = u_{\pm}(x, t \mp \varepsilon x) \mp \varepsilon \tag{4.247}$$

are also solutions; furthermore, they satisfy, in view of eq. (4.243), the initial conditions

$$v_{\pm}(0,t) = \Omega(t) \pm A \operatorname{sech} t. \tag{4.248}$$

Therefore, by Lemma 3, we have

$$|v_{\pm} - \Omega(t/x)| \to 0 \tag{4.249}$$

uniformly in t, as $x \to \infty$; that is, by the definition of $v_{\pm}$,

$$|u_{\pm} - \Omega(t/x \pm \varepsilon) \mp \varepsilon| \to 0, \tag{4.250}$$

in the same limit, uniformly in t. By expanding $\Omega(t/x \pm \varepsilon)$ about $\xi = t/x$ for $x > X > 0$, we can write (4.250) as

$$|u_{\pm} - \Omega(t/x)| < 2(1 + |\eta_0|)\varepsilon \tag{4.251}$$

since $|\mathrm{d}\Omega/\mathrm{d}\xi| \leqslant 2|\eta_0|$, as may easily be checked from eqs. (4.195)–(4.196).

Further, since u lies between u_- and u_+ according to (4.246), we have

$$|u-\Omega(t/x)|<2(1+|\eta_0|)\varepsilon \qquad (4.252)$$

for $x \geqslant X$. The number ε is arbitrary, hence the theorem is proved.

While no estimates relating to asymptotic convergence have been found, it is important to note that the similarity solution here has been shown to be not only an intermediate asymptotic, and stable to initial conditions in this sense, but also stable to strong perturbations of the coefficient $g(x)$, which is merely required to be Hölder continuous and such that $g(x)/x \to \beta$ as $x \to \infty$. Thus, the similarity solution possesses strong global stability.

4.8 The linear (similarity) solution as an intermediate asymptotic – the super-cylindrical Burgers equation

It is known that, if a wave decays sufficiently, its amplitude becomes so small that the nonlinear terms become unimportant. Thus, far away from the origin of the wave, the solution is described by the linearised form of the basic equations. The latter may or may not possess similarity solutions. Moreover, it may happen that the full nonlinear system of equations does not admit similarity solution, while the linearised system does. Thus, the solutions of initial/boundary value problems may tend asymptotically to these linear similarity solutions, perhaps after a much longer time than they would to nonlinear similarity solutions. We shall now discuss one such example (Scott, 1981b). We consider the equation

$$u_x - uu_t = g(x)u_{tt}, \qquad (4.253)$$

where the function $g(x)$ is such that

$$g(x)/x \to \infty \quad \text{as } x \to \infty. \qquad (4.254)$$

This is termed the 'super-cylindrical' case, in contrast to the cylindrical case discussed earlier for which $g(x)/x \to \beta$, a finite number; for example, $g(x)$ is equal to e^x for the spherically symmetric Burgers equation. If we ignore the nonlinear term uu_t in eq. (4.253) and introduce

$$z = \int_0^x g(x)\,dx, \qquad (4.255)$$

then the resulting linear heat equation

$$u_z = u_{tt} \qquad (4.256)$$

has the similarity solution

$$C + D\,\mathrm{erf}(t/2z^{1/2}) \equiv F(x,t) \tag{4.257}$$

which tends to a constant value as z, and x, tend to infinity.

The proof of the intermediate asymptotic nature of $F(x,t)$ with respect to the general initial value problems for (4.253)–(4.254) for a class of functions $u(0,t)$ as initial conditions will be carried out in two stages. First, we show that if $U(x,t)$ is the solution of the heat equation (4.256) with general initial condition $U(0,t) = u_0(t)$ satisfying

$$u_0(t) \to C \pm D \quad \text{as } t \to \pm\infty, \tag{4.258}$$

and $u(x,t)$ is the corresponding solution of the nonlinear problem (4.253–4.254), then $|u - U| \to 0$, as $x \to \infty$, uniformly in t. The second stage comprises a proof that $|U - F| \to 0$, in the same limit, uniformly in t.

Lemma 1. If $u_0(t)$ satisfies (4.258), then $|u - U| \to 0$ as $x \to \infty$, uniformly in t.

Proof. We again write eq. (4.253) as

$$\frac{\partial u}{\partial z} - \frac{\partial^2 u}{\partial t^2} = \frac{1}{G(z)} u \frac{\partial u}{\partial t}, \tag{4.259}$$

where $G(z) = g(x)$, and regard this as an inhomogeneous heat equation with a source term on the right side. The solution of this equation is obtained as

$$\begin{aligned} u(z,t) &= U(z,t) - \int_0^z \int_{-\infty}^{\infty} \frac{1}{2G(\zeta)} \frac{\partial(u^2)}{\partial \tau} \frac{\mathrm{e}^{-(t-\tau)^2/4(z-\zeta)}}{2\pi^{1/2}(z-\zeta)^{1/2}} \mathrm{d}\tau\, \mathrm{d}\zeta \\ &= U(z,t) - \int_0^z \int_{-\infty}^{\infty} \frac{u^2(\zeta,\tau)(t-\tau)\mathrm{e}^{-(t-\tau)^2/4(z-\zeta)}}{G(\zeta)\sqrt{\pi}\cdot[4(z-\zeta)]^{3/2}} \mathrm{d}\tau\, \mathrm{d}\zeta, \end{aligned} \tag{4.260}$$

after an integration by parts with respect to τ. Eq. (4.260) satisfies the initial condition $u(0,t) = U(0,t) = u_0(t)$. By our basic assumptions u is bounded, so we can estimate $|u - U|$:

$$\begin{aligned} |u - U| &\leqslant \int_0^z \int_{-\infty}^{\infty} \frac{u^2(\zeta,\tau)|t-\tau|\mathrm{e}^{-(t-\tau)^2/4(z-\zeta)}}{G(\zeta)\sqrt{\pi}\cdot[4(z-\zeta)]^{3/2}} \mathrm{d}\tau\, \mathrm{d}\zeta \\ &\leqslant K \int_0^z \frac{\mathrm{d}\zeta}{G(\zeta)(z-\zeta)^{1/2}}, \\ &= K \int_0^x \left[\int_\xi^x g(\bar{\xi})\mathrm{d}\bar{\xi} \right]^{-1/2} \mathrm{d}\xi, \end{aligned} \tag{4.261}$$

where $K>0$ is a constant and eq. (4.255) has been used, in addition to the relation $G(z)=g(x)$. In view of (4.254), given $\varepsilon>0$, we can find $X>0$ such that

$$g(x)\geqslant x/\varepsilon \tag{4.262}$$

for any $x\geqslant X$. Therefore, when $x\geqslant X$, we have, from (4.261),

$$\begin{aligned}|u-U|&\leqslant K\left[\int_0^X \frac{(2\varepsilon)^{1/2}}{(x^2-X^2)^{1/2}}\mathrm{d}\xi+\int_X^x \frac{(2\varepsilon)^{1/2}}{(x^2-\xi^2)^{1/2}}\mathrm{d}\xi\right]\\ &=(2\varepsilon)^{1/2}K\left[\frac{X}{(x^2-X^2)^{1/2}}+\frac{\pi}{2}-\arcsin\frac{X}{x}\right].\end{aligned} \tag{4.263}$$

For $x\geqslant 2X$, (4.263) gives

$$|u-U|\leqslant(2\varepsilon)^{1/2}K\left(\frac{1}{\sqrt{3}}+\frac{\pi}{3}\right). \tag{4.264}$$

Since ε is arbitrary, we have proved the lemma.

Lemma 2. $|U-F|\to 0$ as $x\to\infty$, uniformly in t, for any continuous $u_0(t)$ satisfying (4.258).

Proof. The solution of the heat eq. (4.256) with general initial condition $u_0(t)$ satisfying (4.258) is

$$U(x,t)=\int_{-\infty}^{\infty}\frac{u_0(\tau)\mathrm{e}^{-(t-\tau)^2/4z}}{2(\pi z)^{1/2}}\mathrm{d}\tau \tag{4.265}$$

so that

$$U(x,t)-F=\int_{-\infty}^{\infty}\frac{[u_0(\tau)-F(0,\tau)]\mathrm{e}^{-(t-\tau)^2/4\tau}}{2(\pi z)^{1/2}}\mathrm{d}\tau, \tag{4.266}$$

since $F(z,t)$ satisfies the heat equation. The behaviour (4.258) of $u_0(\tau)$ and eq. (4.257) imply that we can find $T>0$ such that

$$|u_0(\tau)-F(0,\tau)|\leqslant\varepsilon$$

for $|\tau|>T$. We now estimate $|U-F|$:

$$\begin{aligned}|U-F|&\leqslant\left(\int_{|\tau|>T}+\int_{|\tau|<T}\right)\frac{|u_0(\tau)-F(0,\tau)|\mathrm{e}^{-(t-\tau)^2/4z}}{2(\pi z)^{1/2}}\mathrm{d}\tau,\\ &\leqslant\varepsilon+\frac{1}{2}\int_{|\tau|<T}|u_0(\tau)-F(0,\tau)|\mathrm{d}\tau(\pi z)^{-1/2},\\ &\leqslant 2\varepsilon\end{aligned} \tag{4.267}$$

for sufficiently large z, that is, for sufficiently large x. This completes the proof.

Combining Lemmas 1 and 2, we find that $|u-F|\to 0$, as $x\to\infty$, uniformly in t. Thus, the solution, in the super-cylindrical case, tends asymptotically to the error function solution of the linearised equation and, for large x, the nonlinear term contributes little to the solution.

4.9 Other generalised Burgers equations

We summarise asymptotic results regarding two other generalised Burgers equations.

(a) The first appears in the physics of ionised gases and is a generalised form of cylindrical Burgers equation (cf. sec. 4.6),

$$u_t = \varepsilon x u_{xx} + (g(x) - u)u_x. \tag{4.268}$$

We consider an ionised gas between two electrodes, which contains ions and electrons with densities $n_i(r)$ and $n_e(r,t)$, respectively, r being the radial distance in a cylindrically symmetric configuration. The ions being heavy and slow are assumed to possess density $n_i(r)$, which does not change with time. The electrons, in comparison, are highly mobile. Referring in particular to the so-called pre-breakdown discharge which spreads out in filamentry form, we wish to find out $n_e(r,t)$ for given $n_i(r)$ and enquire whether, for a given initial (electron) distribution, the electrons stabilise and, if so, to find the time needed for such a stabilisation.

If we write

$$u(x,t) = \int_0^{x^{1/2}} n_e(r,t) r \, dr \tag{4.269}$$

and

$$g(x) = \int_0^{x^{1/2}} n_i(r) r \, dr, \tag{4.270}$$

then it can be shown (Diekmann, Hilhorst & Peletier, 1980, Hilhorst, 1982) that $u(x,t)$ is governed by eq. (4.268) wherein $\varepsilon = 2kT/\mu c_d$, with k, T and μ denoting Boltzman constant, temperature and (electron) mobility, and c_d a fixed constant. The boundary conditions are

$$u(0,t) = 0, \tag{4.271a}$$

$$\int_0^{\infty} [n_i(r) - n_e(r,t)] r \, dr = N > 0,$$

that is,

$$u(\infty, t) = K = g(\infty) - N, \tag{4.271b}$$

where obviously K lies between 0 and $g(\infty)$. The initial condition is

$$u(x, 0) = \psi(x). \tag{4.272}$$

Eqs. (4.268) and (4.271)–(4.272) constitute the mathematical problem. Hilhorst (1982) has proved the existence and uniqueness of the solution to this problem subject to the conditions that g is twice continuously differentiable over $[0, \infty)$, $g(0) = 0, g'(x) > 0$ and $g''(x) < 0$ for all $x \geqslant 0$, and the initial function $\psi(x)$ is continuous, with piecewise continuous derivative on $[0, \infty)$; moreover, $\psi(0) = 0$ and $\psi(\infty) = K \in (0, g(\infty))$. Besides, it is assumed that there exists a constant $M_\psi \geqslant g'(0)$ such that $0 \leqslant \psi'(x) \leqslant M_\psi$ at all points x where ψ' is defined. The corresponding steady problem (with $u_t = 0$ in eq. (4.268)) was investigated in great detail by Diekmann *et al.* (1980), who proved the existence and uniqueness of the solution for this case, and found a matched asymptotic solution. Hilhorst (1982) proved the following theorem regarding the stability of the initial boundary value problem and the evolution of the solution to steady state.

Theorem 1. Let $\Phi(x)$ be the solution to the steady problem.

Suppose ψ satisfies the aforementioned hypotheses, then, for each $x > 0$,

$$\lim_{t \to \infty} u(x, t, \psi) = \Phi(x). \tag{4.273}$$

If $\varepsilon \leqslant g(\infty) - K$, the convergence is uniform on $[0, \infty)$; if $\varepsilon > g(\infty) - K$, it is uniform on all compact intervals of $[0, \infty)$.

The rate of convergence is exponential (in time) if $g(x) \geqslant c_0 x^{1/2}$ for $x \geqslant x_0$, for some positive constants c_0 and x_0; it is algebraic if $\varepsilon < g(\infty) - K$, and the initial function ψ converges algebraically to K as $x \to \infty$.

(b) The second equation concerns unsteady one-dimensional infiltration of water into a homogeneous soil. It has been observed that after sufficient time the moisture profile assumes a certain permanent shape, which then moves downwards with a constant velocity without further change of form. The governing equation for the moisture $u(t, x)$ of the soil, as a function of depth x, measured downward from the ground, and time t is

$$\frac{\partial u}{\partial t} = \frac{\partial}{\partial x}\left[D(u)\frac{\partial u}{\partial x}\right] - \frac{\partial}{\partial x}K(u), \tag{4.274}$$

where $D(u) > 0$, $K(u) > 0$, $D'(u) > 0, K'(u) > 0$, $K''(u) \geqslant \mu > 0$ when $u \geqslant u_0 > 0$.

Depending upon whether there is ground water at depth $x = X$ or not, two initial boundary value problems were posed by Khusnytdinova (1967), and their asymptotic behaviour studied. The intermediate asymptotics or permanent waves were given by the self-propagating solutions $u = U(x - At + c_0)$, where A and c_0 are constant.

Problem 1. Taking into account the initial moisture distribution in the soil and infiltration (of water) at the surface of ground, and with no water at a large depth, the initial and boundary conditions become

$$\left.\begin{aligned} u(t, 0) &= u_1, \quad t > 0, \\ u(0, x) &= u_0(x), \quad 0 \leqslant x < \infty, \\ u_0 &\leqslant u_0(x) \leqslant u_1, \end{aligned}\right\} \tag{4.275}$$

where

$$u_0 = \lim_{x \to \infty} u_0(x)$$

and

$$u_1 = 1$$

at the ground corresponding to full saturation of the soil there.

With regard to eqs. (4.274) and (4.275), the following theorem was proved.

Theorem 2. Let $u(t, x)$ be a solution of the problem (4.274)–(4.275). If the initial function $u_0(x)$ satisfies the condition

$$u_0(x) - u_0 \leqslant M_1 \mathrm{e}^{-\gamma_1 x}, \quad \gamma_1 > \frac{|K'(u_0) - A|}{D(u_0)}, \quad x \geqslant 0, \tag{4.276}$$

where M_1 is a constant, then there exist constants $M > 0$, c_0 and $\beta > 0$, independent of the solution $u(t, x)$, such that

$$|u(t, x) - U(x - At + c_0)| \leqslant M \mathrm{e}^{-\beta t}. \tag{4.277}$$

Problem 2. This problem differs from Problem 1 in that there is water at a finite depth $x = X$ so that the initial and boundary conditions become

$$\begin{aligned} &u(t, 0) = u(t, X) = u_1, \quad u(0, x) = u_0(x), \\ &0 \leqslant x \leqslant X, \quad u_0 \leqslant u_0(x) \leqslant u_1. \end{aligned} \tag{4.278}$$

The corresponding theorem is the following.

Theorem 3. Let $u(t, x)$ be a solution of the boundary value problem (4.274) and (4.278). Then there exist (positive) constants M and β such that

$$|u(t, x) - u_1| < M \mathrm{e}^{-\beta t}. \tag{4.279}$$

5 Numerical solution of nonlinear diffusion equations

5.1 Introduction

It would be apparent from chapters 3 and 4 that analytic methods for nonlinear problems of diffusion (in common with all nonlinear problems) have severe limitations. For example, the non-planar N wave solution obtained by Crighton and Scott (1979) via matched asymptotic expansions, undoubtedly useful in the Taylor-shock régime, holds only over a certain finite time as detailed in chapter 3. The evolutionary shock régime after the Taylor shock and the subsequent decay later to the linear form are scarcely covered by analysis. The self-similar solutions discussed in chapter 4, although the only genuine exact solutions of nonlinear problems (when they exist), are special in nature. They may not exist for a given problem. Even when a self-similar form exists, it satisfies only special initial/boundary conditions so that a given physical problem dictating specific initial and boundary conditions will, in general, not have a self-similar solution. The latter almost always satisfies some singular initial conditions signifying its asymptotic nature. One naturally turns, therefore, to numerical techniques to get a clear qualitative as well as quantitative picture of the phenomena over the entire course of the wave. Since the problems that we discuss involve shocks with discontinuous or steep-fronted initial conditions and have an infinitely long time domain, the numerical methods must be sturdy enough to meet these exigencies. The finite difference or pseudo-spectral methods that may be used for such problems should have other necessary attributes, namely stability, convergence, small truncation error, in addition to economy in computational time in view of the large evolution time of the wave. These requisites have been discussed in a simple and lucid manner by Smith (1978).

In this chapter, we shall discuss two specific numerical approaches to nonlinear diffusion problems: the implicit predictor–corrector method of Douglas and Jones (1963), and the pseudo-spectral approach following mainly the work of Gazdag (1973) and Gazdag and Canosa (1974). Each of these methods enjoys some of the advantages enumerated above, but their selection has been motivated quite considerably by the author's personal experience. Since the finite difference methods are now fairly standard and may be found in several texts such as Richtmyer and Morton (1967), Mitchell and Griffiths (1980), and Jain (1979), we shall content ourselves with a brief description. However, the pseudo-spectral approach has been adopted only recently, so we shall describe it in a more elaborate manner. We choose Fisher's equation and the non-planar Burgers equation as the models for a careful numerical study. The (exact) travelling wave solution of the Fisher's equation and the N wave solution of the plane Burgers equation will provide necessary checks. We may add *en passant* that the travelling wave solution of Fisher's equation will (numerically) be shown to serve as an intermediate asymptotic to which a larger class of solutions with different 'admissible' initial conditions approach in the limit of large time, as explained in chapter 4. We shall also discuss the intermediate asymptotic nature of some self-similar solutions of a GBE with positive or negative damping, and the nonplanar GBE, each with a general power in the nonlinear convection term.

5.2 Implicit finite difference schemes

If we consider the general nonlinear parabolic equation

$$u_{xx} = F(x, t, u, u_x, u_t) \tag{5.1}$$

in the region $(0 \leqslant x \leqslant 1) \times (0 \leqslant t \leqslant T)$, subject to suitable initial and boundary conditions, a straightforward explicit difference scheme would give

$$\frac{1}{h^2}\delta_x^2 u_{i,j} = F\left(ih, jk, u_{i,j}, \delta_x u_{i,j}, \frac{[u_{i,j+1} - u_{i,j}]}{k}\right),$$

$$\delta_x u_{i,j} = \frac{u_{i+1,j} - u_{i-1,j}}{2h},$$

$$\delta_x^2 u_{i,j} = \frac{u_{i+1,j} - 2u_{i,j} + u_{i-1,j}}{2h^2}, \tag{5.2}$$

where $u_{i,j}$ denotes the solution at the discrete point (ih, jk) as obtained by the difference method and δ_x refers to the central space difference. While (5.2) is a convenient formula for obtaining $u_{i,j+1}$ in terms of $u_{i,j}$, simple stability arguments show that this formula severely limits the time step k in terms of the space step h. For example, for the heat equation ($F \equiv u_t$), this restriction is $k \leqslant \frac{1}{2}h^2$ (see Smith (1978)). We note that it is a fortunate circumstance of the finite difference approach to the numerical solution of partial differential equations that the proofs of stability, convergence etc. as applied to linear partial differential equations with constant coefficients carry over with minor changes to those for nonlinear partial differential equations.

In view of the severe restriction on (time) mesh size in the explicit scheme, one is led to search for other methods which may enjoy unconditional stability. Implicit difference methods, which make use of (implicit) information at the current time level as well as the information at the previous time level, belong to such a class. We shall now discuss one such method referred to as the predictor–corrector method. The basic ideas derive from the use of this method for ordinary differential equations. For the first order equation

$$y' = f(x, y) \tag{5.3}$$

the simplest (two-level) predictor–corrector formula is

$$\left.\begin{aligned} w\left(x+\frac{h}{2}\right) &= w(x) + \frac{h}{2} f(x, w(x)), \\ w(x+h) &= w(x) + hf\left(x+\frac{h}{2}, w\left(x+\frac{h}{2}\right)\right). \end{aligned}\right\} \tag{5.4}$$

Here w denotes the discretised value of the function y at the grid points. Under reasonable hypotheses for the initial value problem for eq. (5.3), it is easily seen that the scheme (5.4) is second order accurate, that is, $|y(x) - w(x)| \leqslant Ah^2$ at the grid points. This provides a considerable computational accuracy at the small price of evaluation of w at two points, namely $x + h/2$ and $x + h$.

Douglas and Jones (1963) proposed a finite difference analogue of (5.4) for eq. (5.1) supplemented by the data

$$u(x, 0), \quad u(0, t), \quad u(1, t), \quad 0 < x < 1, \quad 0 < t \leqslant T. \tag{5.5}$$

The intermediate time level $t = (j + \frac{1}{2})k$ predictor

$$\delta_x^2 u_{i,j+1/2} = F(x_i, t_{j+1/2}, u_{i,j}, \delta_x u_{i,j}, \quad [u_{i,j+1/2} - u_{i,j}]/\tfrac{1}{2}k) \tag{5.6}$$

was followed by the corrector

$$\begin{aligned} \tfrac{1}{2}\delta_x^2(u_{i,j+1} + u_{i,j}) = F(x_i, t_{j+1/2}, u_{i,j+1/2}, \\ \tfrac{1}{2}\delta_x[u_{i,j+1} + u_{i,j}], \quad [u_{i,j+1} - u_{i,j}]/k) \end{aligned} \tag{5.7}$$

with the discretely known values in (5.5), $u_{i,0}$, $u_{0,m}$ and $u_{N,m}$, where $0 \leqslant i \leqslant N$, $0 \leqslant m \leqslant M$, $Nh = 1$, and $Mk = T$. Each of eqs. (5.6) and (5.7) involves values of u at three space grid points $(i-1)h$, ih, $(i+1)h$ at its respective time $(j+\frac{1}{2})k$ and $(j+1)k$ so that the evaluation of $u_{i,j+1/2}$ and $u_{i,j+1}$ requires simultaneous solutions of algebraic equations, which take into account the given values $u_{i,0}$, $u_{0,m}$ and $u_{N,m}$. If the function

$$F = f_1(x,t,u)u_t + f_2(x,t,u)u_x + f_3(x,t,u) \tag{5.8}$$

(it covers equations of Burgers type), then it is easily seen that eqs. (5.6) and (5.7) give rise to linear systems of algebraic equations in $u_{i,j+1/2}$ and $u_{i,j+1}$, $0 \leqslant i \leqslant N$. Besides, this system fortunately is tridiagonal so that fast Gauss elimination procedures may conveniently be resorted to (see Smith (1978) and Mitchell and Griffiths (1980)).

Douglas and Jones (1963) have shown that if a solution of eqs. (5.1) and (5.5) exists that has bounded fourth derivatives in $0 < x < 1, 0 < t < T$ and if $F = F(x, t, u_1, u_2, u_3)$ has continuous derivatives with respect to u_i such that

$$\left.\begin{aligned} &\left|\frac{\partial F}{\partial u_i}\right| \leqslant A, \quad i = 1, 2, 3, \\ &\frac{\partial F}{\partial u_3} \geqslant a > 0, \end{aligned}\right\} \tag{5.9}$$

for $0 < x < 1, 0 < t < T, -\infty < u_i < \infty$, then the solution of the predictor–corrector scheme (5.6)–(5.7) converges uniformly to the solution of eq. (5.1) with an error that is $O(h^2 + k^2)$. The predictor–corrector scheme improvises upon the standard Crank–Nicolson scheme (Smith, 1978) in leading to linear algebraic systems of the convenient tridiagonal form. This is in contrast to the latter, which differences nonlinear partial differential equations into a system of nonlinear algebraic equations. Thus the difficult task of solving nonlinear algebraic systems is circumvented, without loss of accuracy of Crank–Nicolson. On the other hand, one could go further and look for three- (or more-) level difference schemes which may be exploited to some other advantage such as a smaller local truncation error or greater stability. This was attempted by Lees (1966). It has been found that this scheme can be used with advantage where steep or discontinuous initial data are involved. Mitchell and Griffiths (1980) consider the nonlinear parabolic equation

$$b(u)u_t = [a(u)u_x]_x, \quad a(u) > 0, \quad b(u) > 0, \tag{5.10}$$

to illustrate Lees' three-level scheme. First, it may be observed that, if eq. (5.10) is approximated by the simplest difference approximation

$$b(u_{i,j})\frac{1}{2k}(u_{i,j+1}-u_{i,j-1})=\delta_x[a(u_{i,j})\delta_x(u_{i,j})], \tag{5.11}$$

with δ_x again denoting the central space difference, we have an unconditionally unstable situation; this was strictly proved for the special case $a(u)=b(u)=1$. However, if eq. (5.11) is first rewritten as

$$b(u_{i,j})(u_{i,j+1}-u_{i,j-1})=2r[a(u_{i+1/2,j})(u_{i+1,j}-u_{i,j}) - a(u_{i-1/2,j})(u_{i,j}-u_{i-1,j})],$$

$$r=\frac{k}{h^2},$$

and then $u_{i+1,j}$, $u_{i,j}$ and $u_{i-1,j}$ are replaced by the averages of values at the three time levels

$$\tfrac{1}{3}(u_{i+1,j+1}+u_{i+1,j}+u_{i+1,j-1}),$$

$$\tfrac{1}{3}(u_{i,j+1}+u_{i,j}+u_{i,j-1}),$$

and

$$\tfrac{1}{3}(u_{i-1,j+1}+u_{i-1,j}+u_{i-1,j-1}),$$

respectively, then we obtain the formula

$$\begin{aligned} b(u_{i,j})(u_{i,j+1}-u_{i,j-1})=\tfrac{2}{3}r\{\alpha^+[(u_{i+1,j+1}-u_{i,j+1})+(u_{i+1,j}-u_{i,j})\\ +(u_{i+1,j-1}-u_{i,j-1})]-\alpha^-[(u_{i,j+1}-u_{i-1,j+1})\\ +(u_{i,j}-u_{i-1,j})+(u_{i,j-1}-u_{i-1,j-1})]\}, \end{aligned} \tag{5.12}$$

where

$$\alpha^+=a\left(\frac{u_{i+1,j}+u_{i,j}}{2}\right),$$

and

$$\alpha^-=a\left(\frac{u_{i,j}+u_{i-1,j}}{2}\right).$$

In writing eq. (5.12), $a(u_{i+1/2,j})$ and $a(u_{i-1/2,j})$ have been replaced by $a([u_{i+1,j}+u_{i,j}]/2)$ and $a([u_{i,j}+u_{i-1,j}]/2)$ so that u need be evaluated only at (spatial) grid points. The system of algebraic equations arising from eq. (5.12) at $t=(j+1)k$ is again linear; however unlike in the two-level case, we need additional data at $t=2k$ by alternative means, besides the initial data, to commence the computation. The data, furthermore, should have an accuracy compatible with that of the three-level scheme. Lees proved the convergence result for the scheme (5.12), namely there exists a constant A, independent of h, k and u such that

$$\max_{i,j} |u_{i,j} - u(ih, jk)| \leqslant A(h^2 + k^2) \tag{5.13}$$

for sufficiently small h and k where $u(ih, jk)$ is the solution of the differential equation (5.10) at the grid point (ih, jk).

In the context of the model equations considered in this monograph, Sachdev and Seebass (1973) used the Douglas–Jones two-level implicit difference scheme for the study of propagation of non-planar N waves. We shall discuss this matter in detail in sec. 5.6 in the light of some more recent calculations. As we remarked in sec. 4.5, Berryman (1977) used Lees' scheme to study the rapid evolution of a separable solution from an arbitrary initial distribution of particles, for the model equation (4.118) describing cross field diffusion in a toroidal octupole plasma-containing device.

Finally, we refer to Meek and Norbury (1982) for some other two-level finite difference schemes for nonlinear parabolic equations.

5.3 Pseudo-spectral numerical scheme

In a pseudo-spectral scheme, the time derivative is differenced according to leap-frog (mid-point rule) or more accurate methods using a Taylor series in time, while the space derivatives are approximated very accurately by means of Fourier transforms. The time is therefore discretised in a forward marching manner while the space derivatives are evaluated to an accuracy permitted by a distribution which can be defined on a finite set of mesh points, by employing a fast Fourier transform (FFT) algorithm (Cooley, Lewis & Welsh, 1969). This scheme has proved very efficient in solving nonlinear model equations, both dispersive and dissipative. For example, Fornberg and Whitham (1978) have adopted this method to solve a whole class of nonlinear dispersive equations typified by the Korteweg–de Vries (K–dV) equation and its generalisations. They studied a variety of problems including solitary wave interactions, wave breaking, the resolution of initial steps and wells, and the development of nonlinear wave train instabilities. Their calculations were performed mainly for graphical use, the time and space step sizes were so chosen as to make all errors in quantities like wave shapes, speeds and positions below the level that can be measured from the graph. The solutions were checked with exact analytic solutions (when they were available) and their accuracy confirmed. Fornberg and Whitham observed that, when time size is halved, the overall error due to time discretisation (by the second-order-accurate leap-frog method) decreases by a factor of 4; in contrast, a halving of the space mesh reduces the spatial discretisation error by several orders of magnitude. The stability analysis for the linearised K–dV equation showed that the stability restriction for the pseudo-spectral approach is $\Delta t/\Delta x^3 < 0.1520$, and 'the limit of (central) finite-difference methods with orders of accuracy increasing to infinity is identical to the pseudo-spectral method using [their] equation (9)' without the nonlinear part. For most practical finite-difference methods, the values of the stability constant are all of the same order of magnitude. The numerical calculations were carried out on the IBM 370/158 in single precision with accuracy between 6 and 7 decimal places, and a real problem with 128 mesh points in the period required approximately 25 milliseconds per time step, independent of the dispersion relation.

In the context of nonlinear diffusion, Gazdag (1973) proposed a scheme which, for time differencing, borrows a feature from the Lax–Wendroff scheme (Richtmyer & Morton, 1967): in the Taylor series in time, the higher time derivatives are replaced by space derivatives with the help of the given (evolutionary) equation. An accurate evaluation of the space derivatives is accomplished by finite Fourier transform. Gazdag refers to this as the accurate space differencing (ASD) method. The accuracy depends on the number of terms retained in the Taylor series for time discretisation, and the extent of the wave number spectrum in the definition of the finite Fourier series. Gazdag gave a stability analysis for a convective equation and computed the travelling shock wave solution of the plane Burgers equation evolving from a discontinuous initial profile. We shall describe this work in detail in the next section.

5.4 Accurate space differencing or pseudo-spectral method for a scalar convective equation

We first discuss the n-dimensional convective equation

$$\frac{\partial \zeta}{\partial t} + \mathbf{v}\cdot\nabla\zeta = 0, \tag{5.14}$$

where $\zeta = \zeta(x, t)$ is a scalar and $\mathbf{v} = (\mathbf{x}, t)$ is a vector $\mathbf{v} = (v_1, v_2, \ldots, v_n)$ which depends on the n spatial co-ordinates $\mathbf{x} = (x_1, x_2, \ldots, x_n)$ and time, and is assumed to be known for the sake of simplicity, although, in general, it would be computed at each time step, for example, by solving the Poisson equation (see Arakawa (1966)). As we remarked earlier, the value of the unknown function ζ^{m+1} at time $t=(m+1)\Delta t$ is found from the values at $t=m\Delta t$ by the Taylor series

$$\zeta^{m+1}=\zeta^m+\frac{\partial\zeta^m}{\partial t}\Delta t+\frac{\partial^2\zeta^m}{\partial t^2}\frac{(\Delta t)^2}{2!}+\cdots+\frac{\partial^p\zeta^m}{\partial t^p}\frac{(\Delta t)^p}{p!}, \tag{5.15}$$

with an error $O((\Delta t)^{p+1})$, where the time derivatives in eq. (5.15) are obtained from eq. (5.14) in terms of space derivatives, by successive differentiation and use of eq. (5.14):

$$\frac{\partial\zeta}{\partial t}=-\mathbf{v}\cdot\nabla\zeta,$$

$$\frac{\partial^2\zeta}{\partial t^2}=-\frac{\partial\mathbf{v}}{\partial t}\cdot\nabla\zeta-\mathbf{v}\cdot\nabla\left(\frac{\partial\zeta}{\partial t}\right),$$

$$\frac{\partial^{l+1}\xi}{\partial t^{l+1}}=-\sum_{i=0}^{l}\frac{l!}{(l-i)!\,i!}\frac{\partial^i\mathbf{v}}{\partial t^i}\cdot\nabla\left(\frac{\partial^{l-i}\zeta}{\partial t^{l-i}}\right) \tag{5.16}$$

for $l=2,3,4,\ldots,p$. The superscript m on ζ has been suppressed in eqs. (5.16) for the sake of convenience. Thus, the Taylor series (5.15) is expressed entirely in terms of ζ^m and its spatial derivatives at $t=m\Delta t$. For one dimension, the latter are found by writing the finite Fourier transform

$$Z(k,t)=\frac{1}{M}\sum_{j=0}^{M-1}\zeta(j\Delta x,t)\,\mathrm{e}^{-ik(j\Delta x)}, \tag{5.17}$$

for

$$k=-\frac{M}{2}+1,-\frac{M}{2}+2,\ldots,\frac{M}{2}-1,$$

$$M=\frac{2\pi}{\Delta x}$$

and its inverse

$$\zeta(j\Delta x,t)=\sum_{k=-M/2+1}^{M/2-1}Z(k,t)\,\mathrm{e}^{ik(j\Delta x)}. \tag{5.18}$$

The spatial derivatives of ζ can be written as

$$\frac{\partial\zeta}{\partial x}(j\Delta x,t)=\sum_{|k|<M/2}Z(k,t)ik\,\mathrm{e}^{ik(j\Delta x)}, \tag{5.19}$$

$$\frac{\partial^2 \zeta}{\partial x^2}(j\Delta x, t) = \sum_{k=-M/2+1}^{M/2} Z(k,t)(-k^2)e^{ik(j\Delta x)}, \tag{5.20}$$

etc. The function $\zeta(x, t)$ is assumed to be some distribution with periodic boundary conditions in the space variable. The period is normalised to be 2π for ease in presentation. This interval is discretised into M equidistant points with spacing $\Delta x = 2\pi/M$, and the function $\zeta(x, t)$ is numerically defined only at these points. If the phenomenon is non-periodic, the (non-normalised) length L is chosen sufficiently large so that all characteristic features (such as soliton interactions or shock evolution) develop in this length.

To study the stability of this scheme, we regard the vector $\mathbf{v}$ as a constant, and enquire whether any single Fourier component, say $e^{i\mathbf{k}\cdot\mathbf{x}}$ at $t = 0$, remains bounded or grows to become unbounded as time increases. In the former case, it is stable, while, in the latter, it is unstable. The solution of eq. (5.14) satisfying $\zeta^0 = \zeta(\mathbf{x}, 0) = e^{i\mathbf{k}\cdot\mathbf{x}}$ is given by

$$\zeta(\mathbf{x}, t) = e^{i\mathbf{k}\cdot(\mathbf{x}-\mathbf{v}t)}. \tag{5.21}$$

The Fourier component $e^{i\mathbf{k}\cdot\mathbf{x}}$ after time Δt becomes

$$\begin{aligned}\zeta(\mathbf{x}, \Delta t) &= e^{i\mathbf{k}\cdot(\mathbf{x}-\mathbf{v}\Delta t)} = e^{i\mathbf{k}\cdot\mathbf{x}} e^{-i\mathbf{k}\cdot\mathbf{v}\Delta t} \\ &= e^{i\mathbf{k}\cdot\mathbf{x}} e^{+i\phi} \equiv \zeta^1 \text{(say)},\end{aligned} \tag{5.22}$$

where

$$\phi = -\mathbf{k}\cdot\mathbf{v}\Delta t$$

so that, in accordance with eq. (5.15),

$$\begin{aligned}\zeta^1 &= \left[1 + i\phi - \frac{\phi^2}{2!} + \cdots + \frac{(i\phi)^p}{p!}\right]\zeta^0 \\ &\equiv \lambda_p(\phi)\zeta^0.\end{aligned} \tag{5.23}$$

Thus, $\lambda_p(\phi) = \sum_{l=0}^{p}[(i\phi)^l/l!]$ is the amplification factor corresponding to p terms in the Taylor series (5.15). It is clear that stability requires $|\lambda_p| < 1$. Gazdag has shown that, with $p = 3, 4, 7$ and 8, this condition is satisfied for $-\pi < \phi < \pi$. The amplification factor $\lambda_p(\phi)$, being a complex quantity, also introduces a phase error

$$\phi_p = \arccos[\mathrm{Re}(\lambda_p)/|\lambda_p|]$$

which may lead to numerical dispersion. Gazdag has shown that, for $p = 3, 4, 7$ and 8, the relative phase error $\delta_p = (\phi_p - \phi)/\phi$ remains small for any given ϕ.

To demonstrate the use and efficacy of the pseudo-spectral approach, Gazdag considered the evolution of the Taylor shock from discontinuous initial conditions under the governance of the plane Burgers equation (cf. sec. 5.6). He gave two approaches; the first requires the split of the equation into convective and diffusive parts, while the second treats the equation as a whole. We skip the first in view of the errors that may accrue from such a split. The plane Burgers equation

$$u_t + uu_x = \frac{\delta}{2} u_{xx} \tag{5.24}$$

with the discontinuous initial conditions

$$u(x,0) = \begin{cases} u_1, & x \leqslant 0, \\ 0, & x > 0, \end{cases} \tag{5.25}$$

was solved by the pseudo-spectral approach to establish in due time the steady state solution

$$u = \frac{1}{2} u_1 \left[1 - \tanh \frac{u_1}{2\delta} \left(x - \frac{u_1 t}{2} \right) \right]. \tag{5.26}$$

The solution $u(t+\Delta t)$ in terms of $u(t)$ was computed from the Taylor series up to $O(\Delta t)^3$:

$$u(t+\Delta t) = u(t) + u_t(t)\Delta t + u_{tt}(t)\frac{(\Delta t)^2}{2!} + u_{ttt}(t)\frac{(\Delta t)^3}{3!}$$

where

$$\left.\begin{aligned} u_t &= -uu_x + \frac{\delta}{2} u_{xx}, \\ u_{tt} &= -u_t u_x - u(u_t)_x + \frac{\delta}{2}(u_t)_{xx}, \\ u_{ttt} &= -u_{tt}u_x - 2u_t(u_t)_x - u(u_{tt})_x + \frac{\delta}{2}(u_{tt})_{xx}. \end{aligned}\right\} \tag{5.27}$$

The right sides in eqs. (5.27) are obtained from the values of the function u at t via the finite Fourier transforms (see (5.17)–(5.20)). The domain of integration

$$D = \{x; 0 \leqslant x \leqslant L\}$$

was partitioned into two subdomains: $D = D_0 + D_1$. The domain

$D_0(0 < x < 0.6)$ had the distribution

$$u(x,t) = \begin{cases} 0, & 0 < x < 0.1, \\ 0.5\{1 - \cos[(x-0.1)\pi/0.3]\} & 0.1 \leqslant x < 0.4, \\ 1, & 0.4 \leqslant x < 0.6, \end{cases} \tag{5.28}$$

for all time during the computation. This 'inert' domain D_0 was chosen to provide a smooth transition between the end points of D_1 as shown in fig. 5.1, to assure periodicity over D, as well as to restrict the expansion of the wave to the left. The data over D_0 (as a part of D) was used for the computation of the space derivatives of u by the Fourier method, but was not updated with time. The computation was advanced in time over the domain D_1. The mesh sizes were chosen to be $\Delta t = 0.001$, $\Delta x = 0.01$. The coefficient δ was chosen to be 0.01. The domain D was represented by 256 mesh points. Fig. 5.1 represents the evolution of the discontinuous profile at successive separations of 0.2 time units. The steady profile emerges at about $t = 0.85$. Table 5.1 shows the comparison of analytic and numerical solutions for $\delta = 0.004$, 0.01 and 0.02. While the overall error is of the order of 0.01% of the maximum value of u_1, and the velocity of the wave profile is 0.4999 (compared to the exact value $\frac{1}{2}$), the forerunner of the shock front shows some deterioration for the smallest (chosen) value of $\delta = 0.004$. The discrepancy, Gazdag avers, may be attributed to the truncation of the spectrum in wave number space so that in the shock

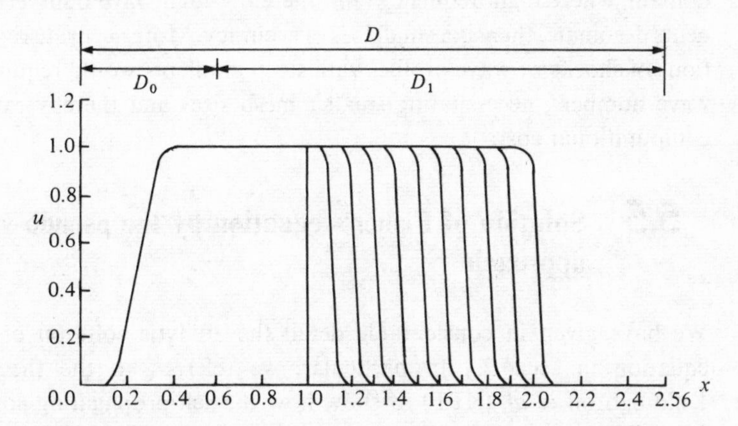

Fig. 5.1. Solution of Burgers equation from discontinuous (step) initial conditions. The time separation between any two consecutive plots is 0.2 units. The domain of integration is also shown (from Gazdag (1973)).

Table 5.1. *Comparison of analytic and numerical results for Fisher's equation for the shock evolved from discontinuous initial profile* (5.28) *under Burgers equation. The steady* (*Taylor*) *shock structure is given for various values of* $\delta = 0.004$, 0.01 *and* 0.02 (*from Gazdag*, 1973).

	$u(x)$					
	$\delta = 0.004$		$\delta = 0.01$		$\delta = 0.02$	
x	Exact	Computed	Exact	Computed	Exact	Computed
1.37	1.000	0.993	0.9975	0.9972	0.9524	0.9524
1.38	1.000	1.005	0.9932	0.9930	0.9239	0.9239
1.39	1.000	0.991	0.9818	0.9814	0.8805	0.8805
1.40	0.999	1.007	0.9520	0.9517	0.8171	0.8172
1.41	0.991	0.979	0.8795	0.8792	0.7305	0.7305
1.42	0.901	0.919	0.7286	0.7285	0.6218	0.6218
1.43	0.428	0.428	0.4969	0.4969	0.4993	0.4993
1.44	0.058	0.062	0.2666	0.2667	0.3769	0.3768
1.45	0.005	0.005	0.1179	0.1180	0.2684	0.2683
1.46	0.000	0.000	0.0469	0.0469	0.1820	0.1819
1.47	0.000	0.000	0.0178	0.0178	0.1189	0.1189
1.48	0.000	0.000	0.0066	0.0066	0.0757	0.0756
1.49	0.000	0.000	0.0024	0.0024	0.0473	0.0472

domain, where high frequency (and therefore high wave number) components dominate, there is a small loss of accuracy. More accurate representations of shocks or wave profiles with steep gradients would require higher wave numbers, necessitating smaller mesh sizes and thereby raising the computational cost.

5.5 Solution of Fisher's equation by the pseudo-spectral approach

We have given in considerable detail the analytic solution of Fisher's equation in sec. 4.2.. In particular, we referred to the theorems of Kolmogoroff *et al.* (KPP) to show how the self-propagating solution of Fisher's equation forms an intermediate asymptotic in the terminology introduced in chapter 4. In the present section we shall verify some of these analytic and stability results numerically using the pseudo-spectral approach, and, in the process, further elucidate the latter method. Here we follow

Gazdag and Canosa (1974). We consider two types of initial value problems for Fisher's equation

$$\left.\begin{aligned} &u_t = u_{xx} + u(1-u), \\ &0 \leqslant u(x,0) \leqslant 1, \end{aligned}\right\} -\infty < x < \infty. \tag{5.29}$$

'Non-local' initial conditions:

$$\lim_{x \to -\infty} u(x,t) = 1, \quad \lim_{x \to +\infty} u(x,t) = 0, \quad t \geqslant 0. \tag{5.30}$$

'Local' initial conditions:

$$\lim_{x \to \pm\infty} u(x,t) = 0, \quad t \geqslant 0. \tag{5.31}$$

Furthermore, it is assumed that all x-derivatives tend to zero as $x \to \pm\infty$. KPP proved the existence and uniqueness of the solution to the non-local problem (5.29)–(5.30). Gazdag and Canosa extended the initial distribution over $0 \leqslant x \leqslant L$ by imposing symmetry about the origin, $u(x,0) = u(-x,0)$, and requiring that the solution u is periodic in space so that $u(x+2nL,t) = u(x,t)$ for all integers n, positive or negative. This arrangement served two purposes. First, both kinds of initial conditions, non-local and local, could be simultaneously treated. The physical explanation for this is that two profiles propagating in opposite directions have no effect on each other, if the distance d_0 from the wave fronts to the origin is large compared to the mean free path, which was taken to be 3 (equal to three times the non-dimensional diffusion coefficient) for the non-dimensional form of the equation. Secondly, the periodicity requirements for the use of finite transform were met. The length L (see fig. 5.2) was chosen sufficiently large so that the profile could evolve from its initial discontinuous shape to its self-propagating form in such a manner that the front identified by the point of inflexion was sufficiently away from the end of the interval L. This was ensured by terminating the numerical experiments as soon as the length d_1 over which u is non-null became rather close to L. Specifically, the number

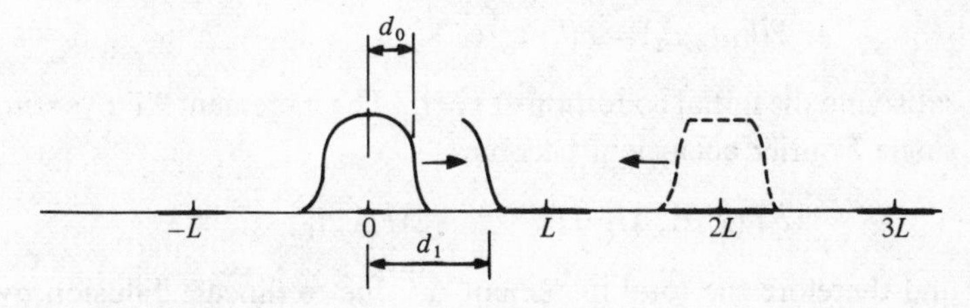

Fig. 5.2. Initial and boundary conditions for Fisher's equation (from Gazdag and Canosa (1974)).

of mesh points chosen was $N = 512$, with mesh spacing $\Delta x = 0.5$ so that L (half the length of the domain of evolution) had 256 points.

Gazdag and Canosa assumed that the nonlinearity and diffusion could be separately treated without serious detriment to the final results. While, as we mentioned earlier, this is strictly not true, the results based on this assumption were found to be satisfactory. Thus, writing

$$\begin{aligned} u(x_n, t_{m+1}) &= u(x_n, t_m) + \Delta u(x_n, t_m, \Delta t) \\ &= u(x_n, t_m) + \Delta f(x_n, t_m, \Delta t) + \Delta g(x_n, t_m, \Delta t) \end{aligned} \tag{5.32}$$

where Δ denotes increment due to the relevant effect in time Δt. The function f is governed by the linear diffusion equation

$$f_t = f_{xx}, \tag{5.33}$$

with initial condition $f(x_n, t_m) = u(x_n, t_m)$, while g, representing the nonlinear local interaction, is governed by the nonlinear first order equation

$$g_t = g(1 - g), \tag{5.34}$$

with initial condition $g(x_n, t_m) = u(x_n, t_m)$. The diffusion equation (5.33) was solved by finite Fourier transform while eq. (5.34) was solved by Taylor series in time. Thus, writing

$$\left.\begin{aligned} f(x_n, t) &= \sum_{j=-N+1}^{N} F(k_j, t)\, e^{ik_j x_n} \\ k_j &= \frac{j\pi}{N\Delta x}, \quad -N < j \leqslant N, \end{aligned}\right\} \tag{5.35}$$

substituting it in eq. (5.33) and equating coefficients of $e^{ik_j x_n}$ on both sides, we have

$$\frac{dF}{dt}(k_j, t) = -k_j^2 F(k_j, t), \quad j = -N+1, \\ -N+2, \ldots, \quad N-1, N,$$

with the solution

$$F(k_j, t_{m+1}) = F(k_j, t_m)\, e^{-k_j^2 \Delta t} \tag{5.36}$$

satisfying the initial condition at $t = t_m$. The increment ΔF over time Δt to a single Fourier coefficient becomes

$$\Delta F(k_j, t_m, \Delta t) = (e^{-k_j^2 \Delta t} - 1) F(k_j, t_m)$$

and therefore the total increment Δf due to (linear) diffusion over Δt is

$$\Delta f(x_n, t_m, \Delta t) = \sum_{j=-N+1}^{N} \Delta F(k_j, t_m, \Delta t)\, e^{ik_j x_n}. \tag{5.37}$$

The contribution due to nonlinearity, Δg, is obtained, say to order $(\Delta t)^4$, by substituting

$$
\begin{aligned}
g_t &= g(1-g), \\
g_{tt} &= (1-2g)g_t, \\
g_{ttt} &= (1-2g)g_{tt} - 2(g_t)^2, \\
g_{tttt} &= (1-2g)g_{ttt} - 6g_{tt}g_t
\end{aligned}
$$

into

$$
\begin{aligned}
\Delta g(x_n, t_m, \Delta t) &= g(x_n, t_{m+1}) - g(x_n, t_m) \\
&= \sum_{i=1}^{4} \frac{\partial^i g(x_n, t_m)}{\partial t^i} \frac{(\Delta t)^i}{i!}.
\end{aligned} \tag{5.38}
$$

The truncation error in the computation of Δg is $O((\Delta t)^5)$. The computations reported by Gazdag and Canosa were carried out with $\Delta t = 0.01$, ensuring a very small truncation error due to discretisation with respect to time. The space discretisations are also of a high degree of accuracy on account of the use of the finite Fourier transforms. Since zero is a solution of the basic equation (5.29), it is obvious that the initial round-off error $\varepsilon(x_n)$ at any point x_n introduced while computing Δu is itself governed by eq. (5.29) (cf. Smith (1978)). It may be expressed as a Fourier series

$$
\varepsilon(x_n) = \sum_{j=-N+1}^{N} E(k_j) e^{ik_j x_n}. \tag{5.39}
$$

The two split effects, nonlinearity and diffusion, influence the round-off error differently. Diffusion smooths out the distribution (5.39). It smears out the rapidly oscillating higher modes (in the shock layer) more effectively. On the other hand, in the tail of the wave, where $u \ll 1$, the local multiplication process (5.34) operates so that $du/dt \approx u$ and $u(t) \approx u_0 e^t$. This implies exponential growth of errors; there would be points x_n in the profile, where

$$
u(x_n) < \varepsilon(x_n) \ll 1.
$$

After some time, the exponential growth of the round-off error at these points will completely overshadow the exact solution $u(x_n)$. Indeed this kind of situation was encountered by Gazdag and Canosa. To obviate it, they (artificially) cut off the tail and set the value of u equal to zero at all points where $|u(x_n)| \leqslant 5 \times 10^{-6}$, an empirically chosen value. Such a procedure does affect the accuracy of the rest of the profile. However, good agreement of their numerical results with the exact solution vindicated this truncation of the tail. We shall revert to the discussion of this matter later in sec. 5.6 in the context of N wave calculations. The formation of the tail and its (unstable) growth are not intrinsic to the numerical method used here. They arise from the particular nature of the solution of the equation. The truncation of the tail is also consistent with the physical nature of the problem since the role of long distance dispersal in the spread of the gene is negligible.

Since the numerical results related to Fisher's equation have a direct bearing on our analysis of sec. 4.2, we consider them in some detail. Fig. 5.3 displays the evolution of the wave of minimum speed with $c = 2$ from the

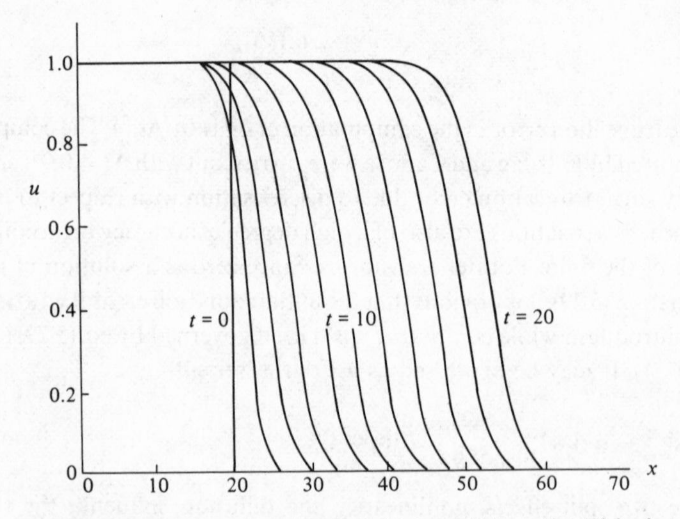

Fig. 5.3. Evolution of step form initial conditions (for Fisher's equation) into a wave with minimum speed $c = 2$. The time separation between any two successive curves is two units (from Gazdag and Canosa (1974)).

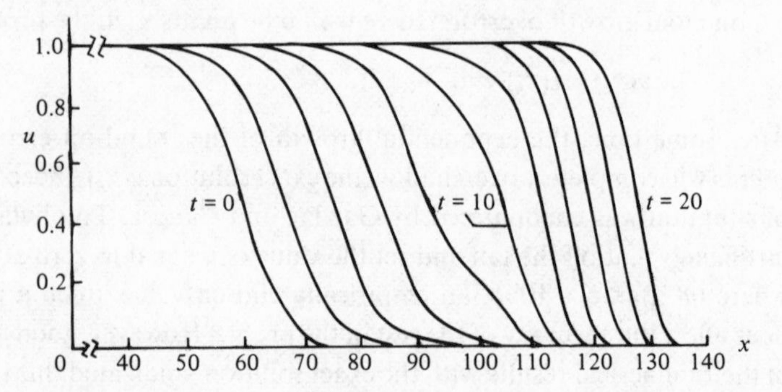

Fig. 5.4. Evolution of the wave of speed $c = 4$ into minimum speed wave (from Gazdag and Canosa (1974)).

KPP initial conditions in the form of a step function. Two successive curves in this figure (and the following) are separated by two dimensionless time units, and the wave speed as measured by the distance travelled by the point of inflexion divided by time was checked to be 2, once the asymptotic self-propagating régime is established after a certain initial evolution. Fig. 5.4 depicts the evolution of an initial self-propagating super-speed wave $u = u(x - ct)$ with $c = 4$. For a certain initial time (about 8 units), the wave propagates unchanged with speed $c = 4$; thereafter, as the figure clearly indicates, the tail of the wave exhibits a clear upward trend or flattening. This is brought about by the truncation of the (theoretically infinitely long) tail on the right, which, cumulatively, changes the form of the wave, leading finally to the emergence of a minimum speed wave with $c = 2$. These computations demonstrate several features:

(i) These waves are stable to arbitrary non-analytic perturbations of finite extent.

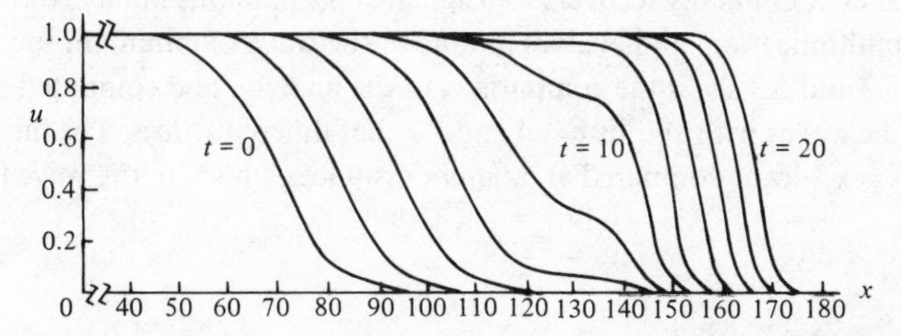

Fig. 5.5. Evolution of the wave of speed $c = 6$ into the minimum speed wave (from Gazdag and Canosa (1974)).

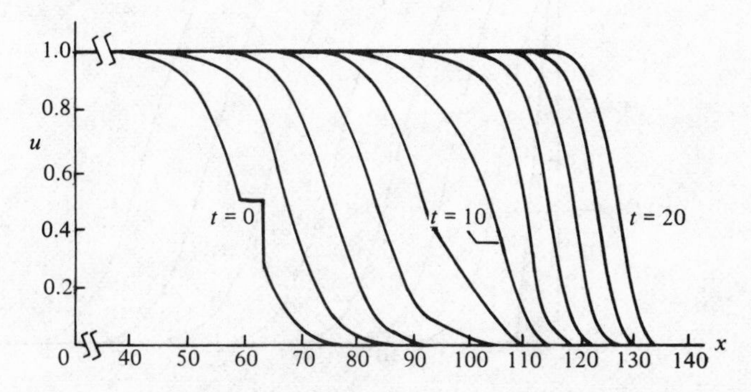

Fig. 5.6. Evolution of a 'dented' wave of speed $c = 4$ into the minimum speed wave (from Gazdag and Canosa (1974)).

(ii) Superspeed waves, after sufficient time, evolve into the wave of minimum speed (see also fig. 5.5 for the evolution of the wave with $c = 6$ into minimum wave speed).

(iii) The truncation of the tail is essential for the stability of numerical results as well as the correct depiction of the physical picture, namely absence of long distance dispersal effect; the evolution time for the minimum speed depends on the cut-off point.

Fig. 5.6 shows an initial wave with $c = 4$ with a finite dent superposed upon it. As the analysis of sec. 4.2 predicted, the perturbation of finite extent, a finite dent, is smoothed out exponentially fast in time, and the third profile (at $t = 4$) is essentially the travelling wave of speed $c = 4$. The later evolution of this wave to one with minimum speed $c = 2$ comes about in the manner described earlier.

Fig. 5.7 shows an initial profile with a (non-analytic) spiky distribution. Diffusion first precipitously smooths it down to a smaller height. Thereafter, it gradually recovers its height ($u = 1$) under the influence of nonlinear multiplication, and evolves in time to the wave of minimum speed. Tables 5.2 and 5.3 show the comparison of the analytic and computed results for the waves with speeds $c = 4$ and $c = 2$ at different times. The function $u(s)$, $s = x - ct$, is compared at different distances which, in the wave frame, are

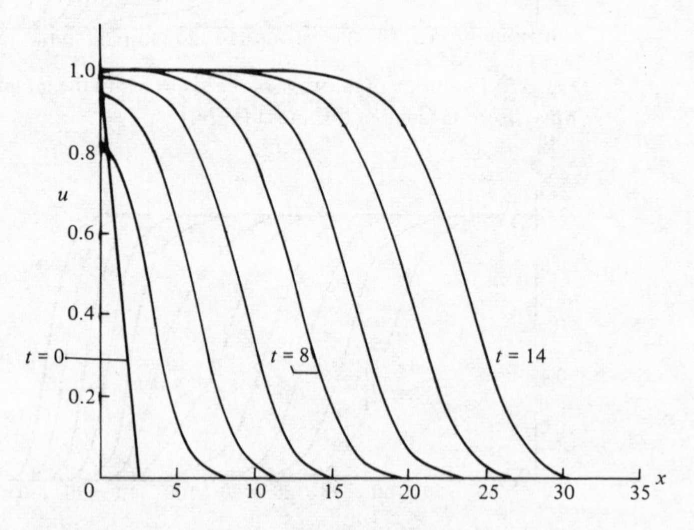

Fig. 5.7. Evolution of a local spiky distribution into the minimum speed wave (from Gazdag and Canosa (1974)).

Table 5.2. *Comparison of analytic (eq.* (4.37)) *and numerical results for Fisher's equation for the wave of speed* $c=4$. *The wave profile* $u(s)$ *computed at* $t=2,4$ *and* 6 *is given as a function of distance, measured in the wave frame, from the point of inflexion of the wave (from Gazdag and Canosa* (1974)).

	u			
s	Analytic	Computed at $t=2$	Computed at $t=4$	Computed at $t=6$
−19.99	0.991	0.991	0.991	0.991
−17.99	0.986	0.986	0.986	0.986
−15.99	0.987	0.987	0.977	0.977
−13.99	0.965	0.965	0.964	0.964
−11.99	0.945	0.945	0.944	0.944
− 9.99	0.914	0.914	0.914	0.913
− 7.99	0.868	0.868	0.868	0.868
− 5.99	0.803	0.803	0.803	0.802
− 3.99	0.715	0.715	0.715	0.714
− 1.99	0.606	0.606	0.606	0.605
0.0063	0.484	0.484	0.483	0.484
2.0063	0.362	0.362	0.361	0.363
4.0063	0.253	0.253	0.253	0.258
6.0063	0.168	0.168	0.168	0.177
8.0063	0.107	0.107	0.107	0.121

measured from the point of inflexion of the wave. The analytic results were derived in sec. 4.2. The time mesh here was chosen to be 0.01.

The excellent agreement between analytic and numerical values even after 600 time steps (of size 0.01) demonstrates the accuracy of the pseudo-spectral (or accurate space differencing) scheme. At $t=6$, there is a perceptible change in the distribution of u near the tail from that given by the analytic solution for $c=4$, reaffirming our earlier comments regarding build-up of changes at the right tail and their propagation to the left resulting ultimately in the emergence of the wave of minimum speed (see Table 5.3 for the evolution of the profile at $t=20$).

Hagstrom and Keller (1986) have treated the boundary conditions for the Fisher's equation more carefully. They find that it is the initial data in the right tail which determines the wave speed. They derive appropriate

Table 5.3. *Comparison of analytic (eq.* (4.37)) *and numerical results for Fisher's equation for the wave of speed* $c = 2$, *similar to Table* 5.2, *at* $t = 20$ (*from Gazdag and Canosa* (1974)).

s	u Analytic	Computed at $t = 20$
− 18.08	0.999	0.999
− 16.08	0.999	0.998
− 14.08	0.998	0.996
− 12.08	0.994	0.992
− 10.08	0.986	0.984
− 8.08	0.967	0.964
− 6.08	0.924	0.923
− 4.08	0.836	0.836
− 2.08	0.678	0.678
− 0.08	0.447	0.447
1.924	0.215	0.214
3.924	0.074	0.069
5.924	0.018	0.016

boundary conditions at the (finite) artificial boundaries, which replace actual infinitely extending ones, by solving linearised problems using Laplace transformation in time. They find that the solutions evolve to a wave of speed $c(\beta)$ provided the initial condition $u_0(x) \sim e^{-\beta x}$ as $x \to \infty$; $c(\beta) = (1 + \beta^2)/\beta, \beta \leqslant 1$, and $c(\beta) = 2, \beta \geqslant 1$.

5.6 Non-planar N wave solution by the implicit and pseudo-spectral finite difference approaches

We observed in secs. 3.4 and 3.7 that matched asymptotic expansions and the generalised similarity solutions of 'a related inverse function' do not solve the non-planar N wave problem in its entirety. While the former gives shock centre as long as the shock is thin, shock displacement due to diffusion is small, and the shock conforms to Taylor structure, the latter gives asymptotic formulae for the Reynolds number for large times. There are clear analytic gaps. There is a time régime when the shock is quite thick (we shall give its quantitative measure presently) for which the asymptotic generalised similarity solution referred to above does not provide even the Reynolds number. A complete analytic solution for the non-planar Burgers equation describing N waves does not seem feasible at present, chiefly because no Hopf–Cole like transformation to linearise this equation has been found. We must, therefore, resort to numerical solution and fill the analytic gaps, verifying, besides, the results with known analytic solutions in different time régimes. This is what we attempt to do in the present section. In the process, we also bring into focus the respective advantages of the implicit finite difference scheme and the pseudo-spectral approach by direct comparison of the numerical results obtained via each method.

We recall that the non-planar Burgers equation is

$$u_t + uu_x + J\frac{u}{2t} = \frac{\delta}{2}u_{xx}, \tag{5.40}$$

where the variables have been defined in secs. 3.4 and 3.7, and $J = 0, 1, 2$ for plane, cylindrical and spherical symmetry, respectively. (A slight change in notation has been introduced for convenience.) Eq. (5.40) is to be solved subject to the initial sawtooth profile of half-length l_0, at $t = t_i$:

$$u(x, t_i) = \begin{Bmatrix} x, & |x| < l_0, \\ 0, & |x| > l_0. \end{Bmatrix} \tag{5.41}$$

This is the time when a steepening of a wave under the lossless equation ($\delta = 0$ in eq. (5.40)) has resulted in (a discontinuous) shock formation. The subsequent evolution of the shock takes place under the competitive influences of nonlinear convection, a small diffusion, and geometrical spreading when $J = 1$ or 2. We have already given a slightly different formulation of this problem due to Crighton and Scott (1979) in sec. 3.4. We shall refer to it again when we need to compare the results. The initial value problem (5.40)–(5.41) differs from that of Sachdev and Seebass (1973) in that the initial profile, in contrast to (5.41), had been endowed with a Taylor structure in the beginning itself, thus missing the embryonic stage altogether. This paper used an implicit finite difference scheme which gave adequate results since a smooth profile presents no serious computational difficulties. However, as pointed out by Crighton (1979), the numerical study by Sachdev and Seebass was not carried out far enough to the linear (old-age) régime. The work by Sachdev, Tikekar and Nair (1986), which we detail now, gives a comprehensive numerical study remedying the short-comings in Sachdev and Seebass (1973). Another contribution of the latter paper was the derivation of the Reynolds number formulae for $J = 1, 2$, by assuming that the slope of the wave profile at the node is given for all time by the inviscid solution

$$u = \begin{cases} \left.\begin{matrix} x/2t, & |x| \leqslant 1, \\ 0, & |x| > 1, \end{matrix}\right\} & J = 1, \\ \left.\begin{matrix} x/t \ln t, & |x| \leqslant 1, \\ 0, & |x| > 1, \end{matrix}\right\} & J = 2. \end{cases} \tag{5.42}$$

Thus, the slope $u_x(0, t)$, according to eq. (5.42), is $1/2t$ and $1/t \ln t$ for $J = 1, 2$, respectively. The lobe Reynolds number of the N wave is defined as

$$R = \frac{1}{\delta} \int_0^\infty u \, \mathrm{d}x. \tag{5.43}$$

Dividing eq. (5.40) by δ and integrating with respect to x from 0 to ∞, we easily derive a first order (ordinary) differential equation for the Reynolds number. The derivative u_x is zero at $x = \infty$ and is obtained from eq. (5.42) at $x = 0$. The first order ordinary differential equation in R is integrated to provide (cf. sec. 3.7)

$$R = \begin{cases} -\dfrac{1}{2} + \left(\dfrac{t_0}{t}\right)^{1/2}, & J = 1, \\ \dfrac{c}{t} - \dfrac{1}{2t} \displaystyle\int_{t_i}^{t} \frac{\mathrm{d}y}{\ln y}, & J = 2, \end{cases} \tag{5.44}$$

where t_0 and c are constants of integration which are evaluated by making use of initial conditions. Thus, for $J = 1$ with $t_i = 1.25$ and $R_i = 23.03$, t_0 was found to be 692.08, while for $J = 2$, with $t_i = 2.61$ and $R_i = 23.03$, the constant c was found to be 60.109. These initial conditions refer to an N wave with Taylor shock structure embedded in it. As we shall see, eq. (5.44) gives a very accurate description of the N wave Reynolds number up to $R \sim 1$. It fails in the low Reynolds number (near-linear and linear) régimes, suggesting the pervasiveness of diffusion in the entire wave profile at this stage so that even the slope at the origin is not given by the inviscid solution (5.42).

Implicit difference scheme

Eq. (5.40), in the difference form according to the Douglas–Jones scheme (5.6)–(5.7), is written out as

$$u_{i+1,j+1/2} - 2(1 + 2h^2/\delta k)u_{i,j+1/2} + u_{i-1,j+1/2}$$
$$= 2u_{i,j}\left[-\frac{2h^2}{\delta k} + \frac{h}{2\delta}(u_{i+1,j} - u_{i-1,j}) + \frac{Jh^2}{2\delta(j+\frac{1}{2})k}\right] \quad \text{(predictor)}, \tag{5.45}$$

$$\left(1 - \frac{h}{\delta}u_{i,j+1/2}\right)u_{i+1,j+1} - 2\left(1 + \frac{2h^2}{k\delta}\right)u_{i,j+1}$$
$$+\left(1 + \frac{h}{\delta}u_{i,j+1/2}\right)u_{i-1,j+1} = \left(-1 + \frac{h}{\delta}u_{i,j+1/2}\right)u_{i+1,j}$$
$$+2\left(1 - \frac{2h^2}{\delta k}\right)u_{i,j} - \frac{h}{\delta}u_{i-1,j}u_{i,j+1/2}$$
$$-u_{i-1,j} + \frac{2Jh^2 u_{i,j+1/2}}{\delta k(j+\frac{1}{2})} \quad \text{(corrector)}. \tag{5.46}$$

Sachdev and Seebass used initial conditions as given by eq. (5.42), supplemented by thin shock layers at the head and the tail of the N wave, wherein the distribution, according to Taylor structure, is

$$u = \frac{u_{\max}}{1 + e^{u_{\max}(x - x_s)/\delta}}. \tag{5.47}$$

Here $u_{\max}$ is the value of u at $x = x_{\max}$ and x_s is the centre of the thin shock such that it is situated half-way between the points where $u = 0.95u_{\max}$ and $u = 0.05u_{\max}$. Assuming these initial values, the solution at the subsequent time is obtained by an iterative scheme that assumes that the solution at a few points on the left of the node of the N wave is anti-symmetric with respect to those on the right. The system of linear algebraic equations at $t_{j+1/2}$ and t_{j+1} is tridiagonal if the value of u at the extreme left-end point can be guessed. This requires changing the value of u at this point at $t_{j+1/2}$ and t_{j+1} until the required anti-symmetry condition is satisfied to the desired accuracy. As the computation proceeds, the wave spreads to the right (and left) and new points have to be added at successive times, where the value of u is significant (say greater than 10^{-6}). As the pulse grows in length, say, becoming twice its original length, the mesh size is increased to keep the matrix of the system of linear equations (5.45)–(5.46) from becoming unwieldy. The calculations were repeated to ensure that the accuracy of the solution did not suffer due to this change of mesh size. First, the numerical scheme was checked with reference to the exact solution of the plane Burgers equation

$$u = \frac{x/t}{1 + e^{x^2/2\delta t}/(e^R - 1)} = \frac{x/t}{1 + (t/t_0)^{1/2} e^{x^2/2\delta t}}. \tag{5.48}$$

The numerical and exact solutions with initial time $t_i = 1$, initial Reynolds number $R_i = 3$, and $t_0 = 364.26$ (cf. eq. (5.48)) were found to compare extremely well. We shall however present some recent numerical results by the implicit scheme as well as those by the pseudo-spectral approach in what follows (Sachdev, Tikekar & Nair, 1986).

Pseudo-spectral approach

As we mentioned earlier, if the initial profile is smooth, the implicit finite difference scheme is adequate and gives good results. The errors do not accumulate and the wave profile is quite accurately determined over its long evolution. In contrast, our experience has shown that a discontinuous initial profile is not properly handled by an implicit scheme. The discontinuous profile does not easily settle down to one with a smooth Taylor shock and the errors introduced are so large that they are likely to vitiate all subsequent computations. To circumvent these difficulties, we had recourse to the pseudo-spectral approach (see sec. 5.3) which proves particularly useful in the early stages of the evolution of the sawtooth profile. Once the steep shock régime has been traversed and the Taylor shock has formed, we switch over to the implicit scheme which then gives good results over the entire evolution of the wave with a computational time which is of the order of one tenth that for the pseudo-spectral approach.

We write the finite Fourier transform

$$\bar{u}(k_j, t) = \frac{1}{K} \sum_{l=0}^{K-1} u(l\Delta x, t) e^{-ik_j l\Delta x} \tag{5.49}$$

and its inverse

$$u(l\Delta x, t) = \sum_{|k_j| < K/2} \bar{u}(k_j, t) e^{ik_j l\Delta x} \tag{5.50}$$

over the interval $(0, 2\pi)$ of x. Here, the spatial mesh size Δx is equal to $2\pi/K$, K denoting the number of mesh points; k_j are wave numbers varying between 0 and $K - 1$. The spatial derivatives easily follow from eq. (5.50):

$$u_x(l\Delta x, t) = \sum_{|k_j| < K/2} (\mathrm{i}k_j)\bar{u}(k_j, t)\,\mathrm{e}^{\mathrm{i}k_j l\Delta x},$$

$$u_{xx}(l\Delta x, t) = \sum_{|k_j| < K/2} (\mathrm{i}k_j)^2\bar{u}(k_j, t)\,\mathrm{e}^{\mathrm{i}k_j l\Delta x}. \tag{5.51}$$

The solution at $(t + \Delta t)$ is obtained from the Taylor series

$$u(x, t + \Delta t) = u(x, t) + (\Delta t)u_t + \frac{(\Delta t)^2}{2!}u_{tt} + \frac{(\Delta t)^3}{3!}u_{ttt} + \cdots,$$

where the time derivatives u_t, u_{tt} etc. are substituted from the basic equation (5.40) in terms of the spatial derivatives as follows:

$$\begin{aligned} u_t &= -uu_x - J\frac{u}{2t} + (\delta/2)u_{xx}, \\ u_{tt} &= -u_t u_x - u(u_t)_x - (J/2t)u_t \\ &\quad + (Ju/2t^2) + (\delta/2)(u_t)_{xx}, \\ u_{ttt} &= -u_{tt}u_x - 2u_{tx}u_t - u(u_{tt})_x - (J/2t)u_{tt} \\ &\quad + (J/t^2)u_t - (Ju/t^3) + (\delta/2)(u_{tt})_{xx}. \end{aligned} \tag{5.52}$$

The spatial derivatives on the right hand sides of eqs. (5.52) were obtained from the distributions of u, u_t, u_{tt} etc. successively, using eqs. (5.51).

The Taylor series for $u(x, t + \Delta t)$ was truncated after four terms so that the error is $O(\Delta t)^4$ (cf. Gazdag (1973)). The initial domain $(0, 2\pi)$ was divided into 256 mesh points in which the (initial) sawtooth profile $u = x$ in $|x| < l_0$ occupied only 80 points so that the profile could grow due to diffusion, as it evolved. We describe the actual calculations with reference to the plane Burgers equation, for which the exact solution is known for all time so that the veracity of the numerical methods can be checked.

Plane Burgers equation

The plane Burgers eq. (5.40) with $J = 0$, subject to initial condition (5.41), was solved using the pseudo-spectral approach. The spatial domain was chosen to be $(0, 2\pi)$ to satisfy periodicity conditions required by the finite Fourier transform. The initial sawtooth was placed in the middle of the domain with adequate space on either side. The mesh sizes were chosen to be $\Delta x = 0.005$ and $\Delta t = 0.01$. Other parameters were specified as follows: $l_0 = 0.205$, $U_0 = u(l_0, t_i) = 0.205$, $\delta = 0.001$ and $t_i = 1$ (see sec. 3.4). The initial

Table 5.4. *'Convergence' of value of* t_0 *as time increases (see eq.* (5.53))

t_n	R_n	$10^{-10}t_0$
1.0	21.354	0.18793
1.1	21.488	0.22534
1.2	21.446	0.22556
1.3	21.406	0.22562
1.4	21.369	0.22564
1.5	21.334	0.22565
1.6	21.302	0.22564
1.7	21.272	0.22563
1.8	21.243	0.22563
1.9	21.216	0.22563
2.0	21.190	0.22563

number of mesh points was taken to be 256. Soon after the computation commenced, a tail of $O(10^{-3})$ was observed which was artificially cut off. Such a spurious tail of a smaller magnitude appeared in the next few time steps, which was again cut off. As the computation proceeded, the tail did not reappear. A similar situation was also encountered by Gazdag and Canosa (1974). We computed the value of the constant t_0 which appears in the solution (5.48), namely

$$t_0 = (e^{R_n} - 1)^2 t_n, \tag{5.53}$$

from the computation of Reynolds numbers R_n at consecutive times $t = t_n$. Table 5.4 shows the convergence of t_0 to a definite finite value before $t \approx 2$. This signalled the end of the 'embryonic shock' regime and emergence of the shock with Taylor structure. At this stage, the exact solution (5.48) becomes applicable. Table 5.5 shows that the numerical solution agrees with exact solution to five decimal places. Table 5.6 shows the values of the Reynolds number, both numerical and analytic as obtained from the formula

$$R = \ln[1 + (t_0/t)^{1/2}], \tag{5.54}$$

from $t = 2$ to $t = 15$. They agree to four decimal places. At about $t \approx 6$, the Taylor shock becomes thick of order $l_0/4$ so that the predictor–corrector implicit scheme was switched on to avoid expensive computation. This scheme gives good results for an initial profile with sufficiently smooth

Table 5.5. *Comparison of numerical solutions of Burgers' equation, computed by PS* and IMP* methods, with exact solution at* $t = 3$ *(smooth initial profile at* $t = 2$ *was obtained from the discontinuous one at* $t = 1$ *using PS method).*

x	u_{IMP}	u_{PS}	Exact	Difference	
(1)	(2)	(3)	(4)	(2)–(4)	(3)–(4)
0.01	0.00333	0.00333	0.00333	0.0	0.0
0.11	0.03667	0.03667	0.03667	0.0	0.0
0.21	0.07	0.07	0.07	0.0	0.0
0.23	0.07667	0.07667	0.07667	0.0	0.0
0.25	0.08333	0.08333	0.08333	0.0	0.0
0.27	0.09	0.08999	0.08999	0.00001	0.0
0.29	0.09661	0.09658	0.09658	0.00003	0.0
0.31	0.10283	0.10263	0.10262	0.00021	0.00001
0.33	0.10479	0.10392	0.10392	0.00087	0.0
0.35	0.07413	0.07454	0.07455	0.00042	−0.00001
0.37	0.01630	0.01704	0.01707	−0.00077	−0.00003
0.39	0.00165	0.00163	0.00164	0.00001	−0.00001
0.41	0.00013	0.00012	0.00012	0.00001	0.0
0.43	0.00001	0.00001	0.00001	0.0	0.0

Table 5.5 (contd.). *Solution of Burgers' equation at* $t = 15$ *(smooth initial profile at* $t = 14$ *was obtained from the discontinuous profile at* $t = 1$ *using PS method).*

x	u_{IMP}	u_{PS}	Exact	Difference	
(1)	(2)	(3)	(4)	(2)–(4)	(3)–(4)
0.03	0.002	0.002	0.002	0.0	0.0
0.33	0.022	0.022	0.022	0.0	0.0
0.63	0.04196	0.04196	0.04196	0.0	0.0
0.66	0.04386	0.04385	0.04385	0.00001	0.0
0.69	0.04543	0.04539	0.04539	0.00004	0.0
0.72	0.04562	0.04550	0.04550	0.00012	0.0
0.75	0.04049	0.04037	0.04037	0.00012	0.0
0.78	0.02453	0.02474	0.02474	0.00021	0.0
0.81	0.00833	0.00843	0.00843	−0.00010	0.0
0.84	0.00194	0.00192	0.00192	0.00002	0.0
0.87	0.00038	0.00037	0.00037	0.00001	0.0
0.9	0.00007	0.00007	0.00007	0.0	0.0

*PS stands for pseudo-spectral method while IMP denotes (predictor–corrector) implicit scheme.

Table 5.6. *Comparison of the exact Reynolds numbers (eq. (5.54)) with those obtained numerically using PS method for the plane Burgers equation. The table also gives the slope* $u_x(0, t)$.

			Slope	
t	R (Num)	R (Exact)	Num	Exact
2.0	21.19044	21.19044	0.50000	0.50000
3.0	20.98770	20.98771	0.33334	0.33333
5.0	20.73229	20.73230	0.20000	0.20000
7.0	20.56402	20.56406	0.14286	0.14286
9.0	20.43838	20.43840	0.11111	0.11111
11.0	20.33804	20.33806	0.09091	0.09091
13.0	20.25451	20.25454	0.07692	0.07692
15.0	20.18295	20.18299	0.06667	0.06667

shocks. Table 5.5 confirms this accuracy by comparison with the exact solution and the results of the pseudo-spectral approach at several time instants. Table 5.5 shows that the implicit scheme at an early time, say $t = 3$, has an error $O(9 \times 10^{-4})$ in the shock layer while the pseudo-spectral one has an error $O(3 \times 10^{-5})$. As the profile becomes smoother, say at $t = 6$, the error by the pseudo-spectral approach reduces to $O(10^{-6})$ while the implicit scheme has still an error $O(3 \times 10^{-4})$ in the shock layer but a smaller one $O(10^{-6})$ elsewhere. This inaccuracy in the implicit scheme in the steep shock layer seems intrinsic to it, but, fortunately, the error does not build up as time progresses, as evidenced by the solutions at $t = 3$ and $t = 15$ (see Table 5.5) and those at much later times (see Table 5.7). These tables show the accuracy of the implicit scheme in the large Reynolds number régime. The smaller Reynolds number régime, say, of order 3 or less is covered by the implicit scheme much more accurately (see Sachdev and Seebass (1973)). It is interesting to note that refining the mesh size leads to a much higher accuracy for the pseudo-spectral approach, but does not improve the accuracy of the implicit scheme in the shock layer (see Table 5.8). As the wave evolves, the non-vanishing part of the profile occupies a larger domain. At $t = 100$, for example, there are 1200 mesh points in the non-zero part of the profile. To conserve the computational expense, the mesh size was doubled several times to be able to do with 256 points only. The accuracy was, however, ensured by reference to the exact solution. For non-planar cases for which no exact solution is available, the

Table 5.7. *Numerical (IMP) and exact solutions of plane Burgers equation.*

$u(x, 10)$			$u(x, 50)$			$u(x, 100)$		
x	Num	Exact	x	Num	Exact	x	Num	Exact
0.005	0.00050	0.00050	0.02	0.00040	0.00040	0.04	0.00040	0.00040
0.05	0.00500	0.00500	0.12	0.00240	0.00240	0.24	0.00240	0.00240
0.10	0.01000	0.01000	0.22	0.00440	0.00440	0.44	0.00440	0.00440
0.20	0.02000	0.02000	0.32	0.00640	0.00640	0.64	0.00640	0.00640
0.40	0.04000	0.04000	0.42	0.00840	0.00840	0.84	0.00840	0.00840
0.5	0.05000	0.05000	0.52	0.01040	0.01040	1.04	0.01040	0.01040
0.55	0.05475	0.05475	0.62	0.01240	0.01240	1.24	0.01240	0.01240
0.56	0.05555	0.05550	0.72	0.01440	0.01440	1.44	0.01440	0.01440
0.57	0.05618	0.05611	0.92	0.01840	0.01840	1.64	0.01638	0.01635
0.58	0.05650	0.05640	1.02	0.02040	0.02040	1.76	0.01734	0.01719
0.59	0.05628	0.05615	1.12	0.02238	0.02238	1.80	0.01717	0.01740
0.60	0.05510	0.05494	1.22	0.02420	0.02418	1.82	0.01703	0.01673
0.65	0.02076	0.02100	1.28	0.02466	0.02466	2.04	0.00325	0.00350
0.7	0.00112	0.00112	1.32	0.02377	0.02366	2.24	0.00007	0.00006

Table 5.7 (contd.)

$u(x, 200)$			$u(x, 400)$			$u(x, 500)$		
x	Num	Exact	x	Num	Exact	x	Num	Exact
0.04	0.00020	0.00020	0.08	0.00020	0.00020	0.08	0.00016	0.00016
0.24	0.00120	0.00120	0.48	0.00120	0.00120	0.88	0.00173	0.00176
0.44	0.00220	0.00220	0.88	0.00220	0.00220	1.68	0.00330	0.00336
0.64	0.00320	0.00320	1.28	0.00320	0.00320	2.48	0.00487	0.00496
0.84	0.00420	0.00420	1.68	0.00420	0.00420	3.28	0.00644	0.00656
1.04	0.00520	0.00520	2.08	0.00520	0.00520	3.92	0.00743	0.00749
1.24	0.00620	0.00620	2.48	0.00620	0.00620	4.00	0.00732	0.00735
1.44	0.00720	0.00720	2.88	0.00720	0.00720	4.08	0.00695	0.00699
1.64	0.00820	0.00820	3.28	0.00817	0.00817	4.48	0.00133	0.00145
1.84	0.00920	0.00920	3.44	0.00847	0.00840	4.88	0.00005	0.00005
2.04	0.01020	0.01020	3.52	0.00851	0.00840			
2.24	0.01119	0.01118	3.60	0.00835	0.00821			
2.48	0.01211	0.01204	4.08	0.00090	0.00096			
2.52	0.01210	0.01201	4.48	0.00002	0.00002			
2.64	0.01080	0.01072						
2.84	0.00297	0.00310						
3.04	0.00022	0.00022						
3.24	0.00001	0.00001						

Table 5.8. *Solution of plane Burgers equation at $t = 3$, obtained by using different mesh sizes and different methods. Initial profile at $t = 2$ was taken to be smooth.*

x (1)	u_{IMP} $\Delta x = 0.01$ (2)	u_{IMP} $\Delta x = 0.005$ (3)	u_{PS} $\Delta x = 0.01$ (4)	u_{PS} $\Delta x = 0.005$ (5)	u_{Exact} (6)
0.01	0.0033334	0.0033334	0.003333	0.0033334	0.0033333
0.03	0.0100002	0.0100002	0.0100007	0.0100002	0.01
0.05	0.0166670	0.0166670	0.0166678	0.0166669	0.0166667
0.07	0.0233338	0.0233338	0.0233350	0.0233337	0.0233333
0.09	0.0300006	0.0300006	0.0300021	0.0300005	0.03
0.11	0.0366674	0.0366675	0.0366693	0.0366673	0.0366667
0.13	0.0433342	0.0433343	0.0433365	0.0433341	0.0433333
0.15	0.0500010	0.0500010	0.0500037	0.0500009	0.05
0.17	0.0566678	0.0566679	0.0566710	0.0566677	0.0566667
0.19	0.0633347	0.0633348	0.0633383	0.0633345	0.0633333
0.21	0.0700015	0.0700016	0.0700057	0.0700013	0.0699999
0.23	0.0766683	0.0766684	0.0766730	0.0766679	0.0766663
0.25	0.0833348	0.0833343	0.0833390	0.0833331	0.0833312
0.27	0.0900003	0.0899947	0.0899963	0.0899895	0.0899869
0.29	0.0966602	0.0966128	0.0965877	0.0965800	0.0965760
0.31	0.1033360	0.1028331	0.1026357	0.1026284	0.1026216
0.33	0.1084517	0.1047879	0.1039273	0.1039248	0.1039148
0.35	0.0712843	0.0741339	0.0745553	0.0745409	0.0745504
0.37	0.0144973	0.0163013	0.0170474	0.0170449	0.0170647
0.39	0.0016943	0.0016499	0.0016279	0.0016321	0.0016365

accuracy was confirmed by occasionally repeating the computations with both refined and crude mesh sizes (see Table 5.9 for the cylindrical case).

To compare the two numerical schemes, we also treated the evolution of the discontinuous initial profile by the implicit scheme. It was observed that the error in the numerical solution was $O(10^{-3})$; besides, the value of the parameter t_0 would oscillate and not converge for a long time.

The position of the shock centre X, according to Crighton and Scott (1979), is

$$l_0 x = X = l_0 T^{1/2}(1 - \varepsilon \ln T) \tag{5.55}$$

where

$$T = 1 + \frac{\gamma + 1}{2} U_0(t - t_i)/l_0, \quad \varepsilon = \delta/(\gamma + 1)U_0 l_0, \tag{5.56}$$

Table 5.9. *Effect of change of mesh size on the solution for the cylindrical Burgers equation. The table shows the solutions at $t = 100$, obtained by using $\Delta x = 0.005$ and $\Delta x = 0.02$. The initial profile at $t = 1$ was taken to be discontinuous.*

x (1)	$u(x, 100)$ $\Delta x = 0.005$ (2)	$^*u(x, 100)$ $\Delta x = 0.02$ (3)	Difference (2)–(3) (4)
0.08	0.000356	0.000356	0.0
0.16	0.000700	0.000700	0.0
0.24	0.001017	0.001017	0.0
0.32	0.001290	0.001291	−0.000001
0.40	0.001500	0.001497	0.000003
0.48	0.001612	0.001612	0.0
0.56	0.001613	0.001613	0.0
0.64	0.001492	0.001492	0.0
0.72	0.001266	0.001265	0.000001
0.80	0.000980	0.000978	0.000002
0.88	0.000689	0.000688	0.000001
0.96	0.000442	0.000441	0.000001
1.04	0.000259	0.000258	0.000001
1.12	0.000140	0.000139	0.000001
1.20	0.000070	0.000070	0.0

*The initial mesh size $\Delta x = 0.005$ was doubled twice.

and $\gamma = 1.4$ (see sec. 3.4 for definitions of U_0, l_0 etc.). The numerical shock centre (see Lighthill (1956)) is approximated by

$$X = (X_1 + X_2)/2, \tag{5.57}$$

where X_1 and X_2 are spatial co-ordinates of the points with $u = 0.95u_{max}$ and $u = 0.05u_{max}$, respectively, and u_{max} is the maximum amplitude of the wave. The values of the shock centre thus obtained differ from eq. (5.55) by less than 7% up to $t \approx 30$. Table 5.10 compares these results and also contains the shock width

$$S = X_2 - X_1. \tag{5.58}$$

The shock width at $t = 500$ is about 2.9 times the initial (half) length of the wave profile. Table 5.10 also includes the maximum of the difference between the numerical shock structure and the Taylor shock structure given by eq. (5.47).

Table 5.10. *Shock details for plane Burgers equation,* $t_i = 1$, $\varepsilon = 0.0099$ *(see eq.* (5.55)).
x_{max} = *location of the point of maximum amplitude* u_{max}.
u_d = *maximum of the difference between Taylor-shock structure and numerical shock structure.*
CS – Crighton and Scott (1979).

t	x_{max}	u_{max}	Shock width	Shock centre Numerical	Shock centre CS	Difference (5)–(6)	u_d	u_d/u_{max}
(1)	(2)	(3)	(4)	(5)	(6)	(7)	(8)	(9)
1.1	0.2	0.180	0.020	0.218	0.217	0.001	0.027	0.151
1.3	0.216	0.162	0.027	0.238	0.238	0.0	0.014	0.088
1.5	0.232	0.150	0.033	0.254	0.258	−0.004	0.025	0.169
2.0	0.267	0.129	0.039	0.292	0.302	−0.01	0.04	0.309
2.5	0.296	0.115	0.04	0.328	0.340	−0.012	0.045	0.392
3.0	0.324	0.105	0.045	0.358	0.373	−0.015	0.046	0.442
3.5	0.35	0.097	0.048	0.385	0.404	−0.018	0.046	0.478
4.0	0.373	0.091	0.051	0.413	0.433	−0.02	0.045	0.501
5.0	0.416	0.081	0.056	0.460	0.485	−0.025	0.044	0.54
10.0	0.582	0.057	0.08	0.646	0.687	−0.041	0.034	0.601
15.0	0.708	0.046	0.099	0.787	0.840	−0.053	0.028	0.618
20.0	0.81	0.04	0.114	0.906	0.969	−0.063	0.025	0.623
30.0	0.992	0.032	0.14	1.104	1.183	−0.079	0.021	0.638
50.0	1.273	0.025	0.18	1.415	1.521	−0.106	0.016	0.644

Table 5.11. *Reynolds numbers for plane Burgers equation at different times as obtained from numerical (IMP) and exact solutions. This table also gives* $u_x(0, t)$.

t	R Numerical	R Exact	$u_x(0,t)$ Numerical	$u_x(0,t)$ Exact
2.0	21.19044	21.19044	0.50000	0.50000
3.0	20.98770	20.98771	0.33334	0.33333
5.0	20.73211	20.73230	0.20000	0.20000
10.0	20.38545	20.38572	0.10000	0.10000
15.0	20.18287	20.18299	0.06667	0.06667
20.0	20.03925	20.03915	0.05000	0.05000
50.0	19.58110	19.58100	0.02000	0.02000
100.0	19.23382	19.23444	0.01000	0.01000
200.0	18.88652	18.88785	0.00500	0.00500
300.0	18.68374	18.68512	0.00333	0.00333
400.0	18.53919	18.54127	0.00250	0.00250

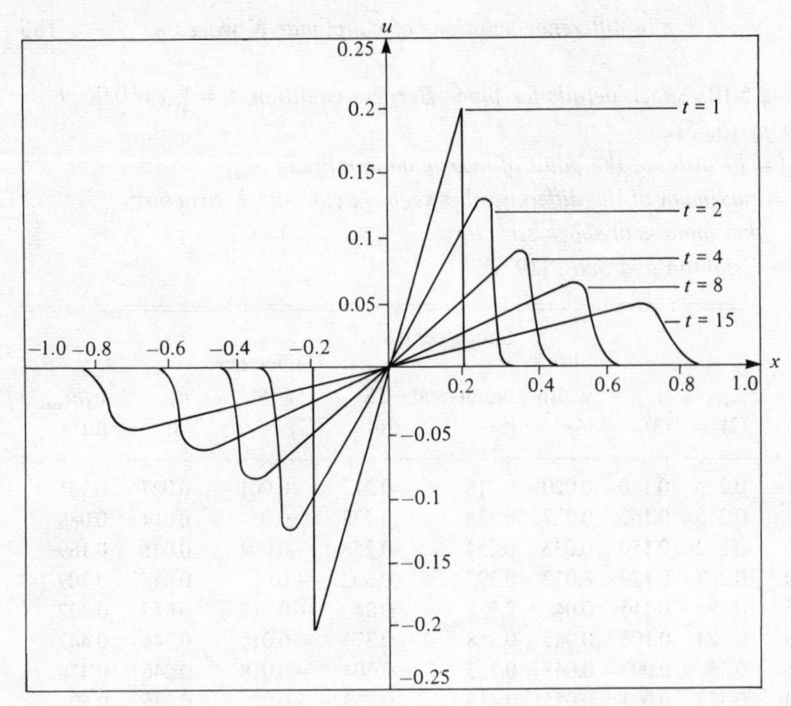

Fig. 5.8. *N* wave solution of plane Burgers equation: embryo shock to Taylor shock.

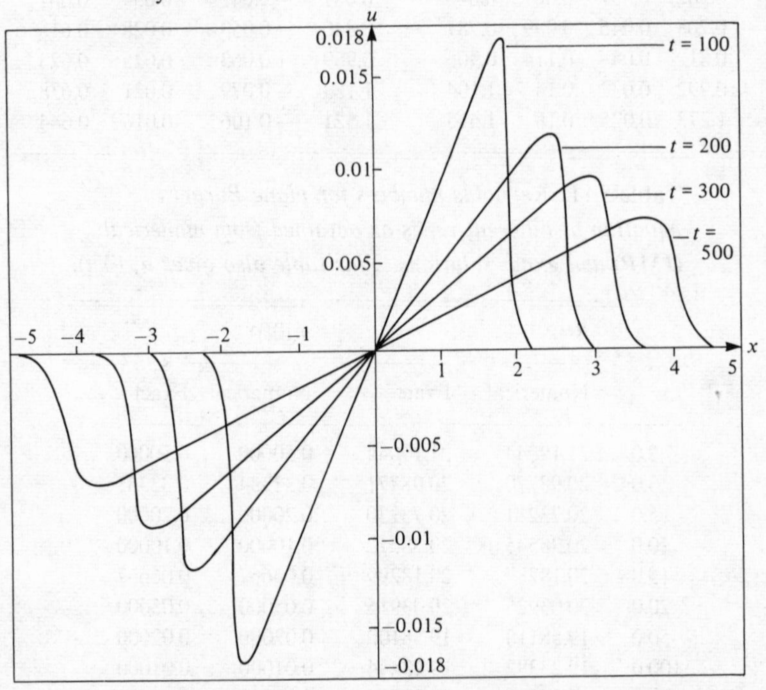

Fig. 5.9. *N* wave solution of plane Burgers equation: thick shock régime.

The small discrepancy here confirms that, in this high Reynolds number régime, the shock evolves according to Taylor structure. The values of the Reynolds number, analytic and numerical, agree to five significant places (see Table 5.11). The slope $u_x(0, t)$ is also given in Table 5.11 and agrees very closely with that according to the inviscid solution $u = x/t$.

The plane N wave decays very slowly; the value of the lobe Reynolds number changes from 21 at $t = 1$ to 18.1 at $t = 500$. The verification of the numerical results in the high Reynolds number régime was terminated at this stage. The diffusion of the plane N wave is shown in figs. 5.8 and 5.9.

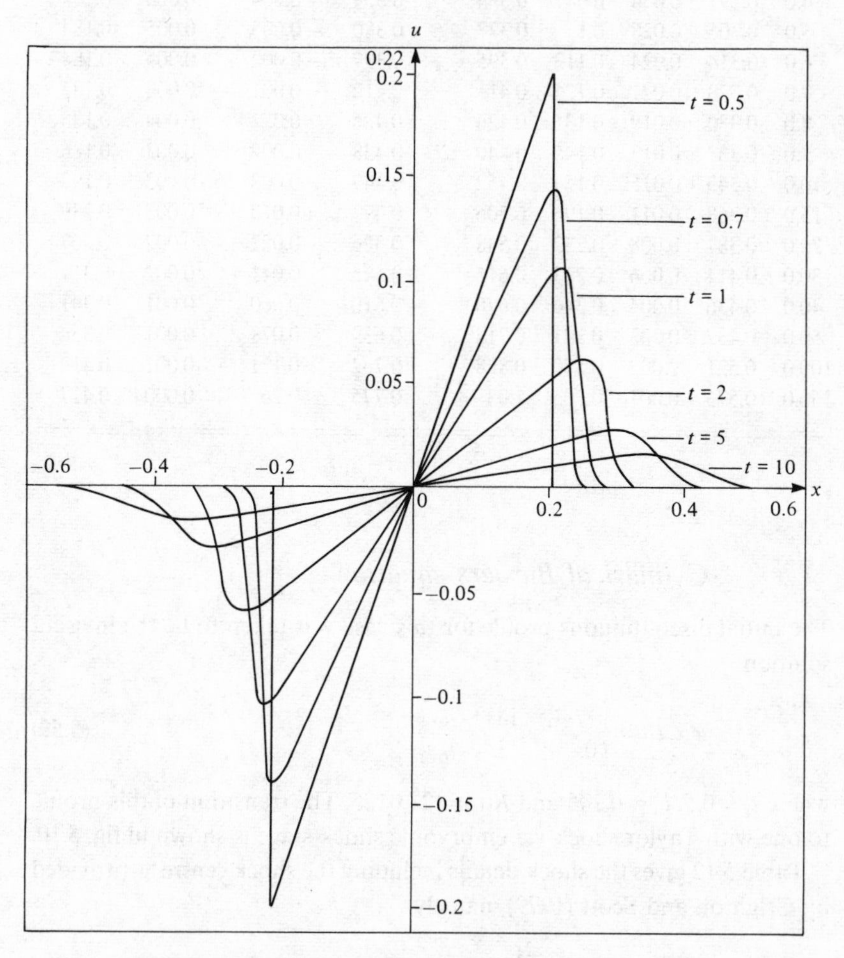

Fig. 5.10. N wave solution of cylindrical Burgers equation: embryo shock to Taylor shock.

Table 5.12. *Shock details for cylindrical Burgers equation, $T_0 = 1.2$, $\varepsilon = 0.0165$ (see eq. (5.60)); abbreviations are explained in Table 5.10*

t (1)	x_{max} (2)	u_{max} (3)	Shock width (4)	Shock centre Numerical (5)	Shock centre CS (6)	Difference (5)–(6) (7)	u_d (8)	u_d/u_{max} (9)
1.0	0.221	0.108	0.035	0.249	0.250	−0.001	0.013	0.118
2.0	0.255	0.061	0.057	0.297	0.301	−0.004	0.011	0.176
3.0	0.275	0.044	0.073	0.330	0.334	−0.004	0.008	0.182
4.0	0.291	0.034	0.087	0.355	0.359	−0.004	0.006	0.184
5.0	0.305	0.028	0.1	0.377	0.380	−0.003	0.005	0.184
6.0	0.314	0.024	0.113	0.395	0.397	−0.002	0.004	0.184
7.0	0.323	0.021	0.125	0.411	0.412	−0.001	0.004	0.185
8.0	0.330	0.019	0.135	0.426	0.426	0.0	0.004	0.185
9.0	0.337	0.017	0.145	0.440	0.438	0.002	0.003	0.186
10.0	0.343	0.015	0.154	0.452	0.449	0.003	0.003	0.197
15.0	0.368	0.011	0.196	0.506	0.493	0.013	0.003	0.239
20.0	0.387	0.008	0.232	0.548	0.526	0.022	0.002	0.267
30.0	0.414	0.006	0.293	0.615	0.575	0.041	0.002	0.311
40.0	0.436	0.004	0.346	0.670	0.610	0.060	0.001	0.341
50.0	0.437	0.003	0.391	0.717	0.639	0.078	0.001	0.354
100.0	0.521	0.002	0.570	0.898	0.762	0.171	0.001	0.412
150.0	0.573	0.001	0.705	1.03	0.775	0.26	0.0004	0.427

Cylindrical Burgers equation

The initial discontinuous profile for this case was taken to be the inviscid solution

$$u(x, t_i) = \begin{Bmatrix} x/2t_i, & |x| < l_0, \\ 0, & |x| > l_0, \end{Bmatrix} \tag{5.59}$$

where $t_i = 0.5$, $l_0 = 0.205$ and $R(t_i) = 21.0125$. The transition of this profile to one with Taylor shock via embryonic shock stage is shown in fig. 5.10.

Table 5.12 gives the shock details including the shock centre as provided by Crighton and Scott (1979), namely

$$l_0 x = X = l_0 T^{1/2} \left\{ 1 - \frac{\varepsilon}{2}[T - 1 + (T_0 - 1) \ln T] \right\}, \tag{5.60}$$

where

$$\varepsilon = 2\delta/(\gamma + 1)U_0 l_0 T_0,$$
$$T = 1 + (\gamma + 1)U_0(t_i^{1/2}t^{1/2} - t_i)/l_0$$

and

$$T_0 = \frac{\gamma + 1}{l_0} U_0 t_i.$$

We take $U_0 = l_0$ and $\delta = 0.001$ for $j = 1, 2$ just as for $j = 0$. The numerical shock centre agrees with eq. (5.60) up to $t \approx 25$, with an error less than 5%. This is the stage when the shock becomes rather thick, approximately $1.4l_0$, long after the assumption of a thin shock underlying the analysis of matched asymptotic expansions has broken down. It is interesting that the matched asymptotic expansion gives good results thus far. The computations were carried from $t = 0.5$ to $t = 2000$. Figs. 5.10–5.12 show the evolution of the wave in three typical time régimes – embryo/Taylor shock, thick shock and old age. The Reynolds numbers – numerical, analytic (see

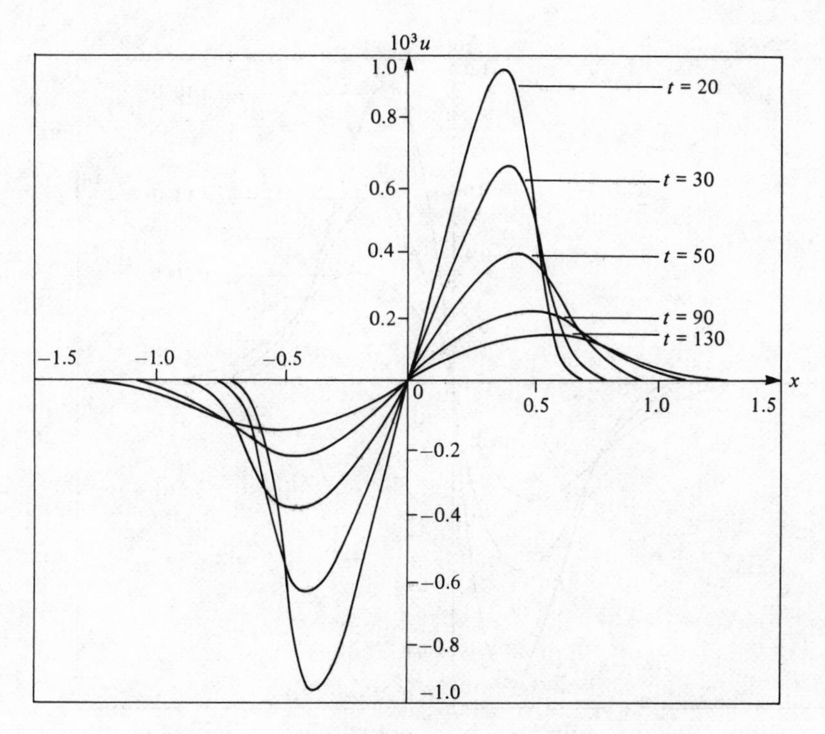

Fig. 5.11. *N* wave solution of cylindrical Burgers equation: thick shock régime.

eq. (5.61) below) and by eq. (5.44) are given in Table 5.13. This table also contains the slope $u_x(0,t) = 1/2t$ as calculated from the inviscid solution as well as from the numerical solution. The two agree to three decimal places up to $t \approx 100$ when $R \sim 1$. Over this duration, the numerical value of the Reynolds number also agrees with that by eq. (5.44). After this time, eq. (5.44) begins to fail so that at $t \approx 1000$, eq. (5.44) gives a negative value of the Reynolds number! Thus, the validity of eq. (5.44) is limited to higher Reynolds numbers $R \geqslant 1$ only.

The generalised similarity solution of the 'inverse function' in sec. 3.7 led to the following expression for Reynolds number:

$$R = (\tilde{c}/t)^{1/2} + (1/tb_3)^{1/2} \ln (t^{1/2} + 2/b_3^{1/2})/(t^{1/2} + 1/b_3^{1/2}). \qquad (5.61)$$

Table 5.13 shows that eq. (5.61) gives good results for $t > 300$ and the accuracy improves with time, confirming its asymptotic nature. The values of the constants in eq. (5.61) were determined by matching it with the numerical solution at $t = 500$ and $t = 1600$. These were found to be

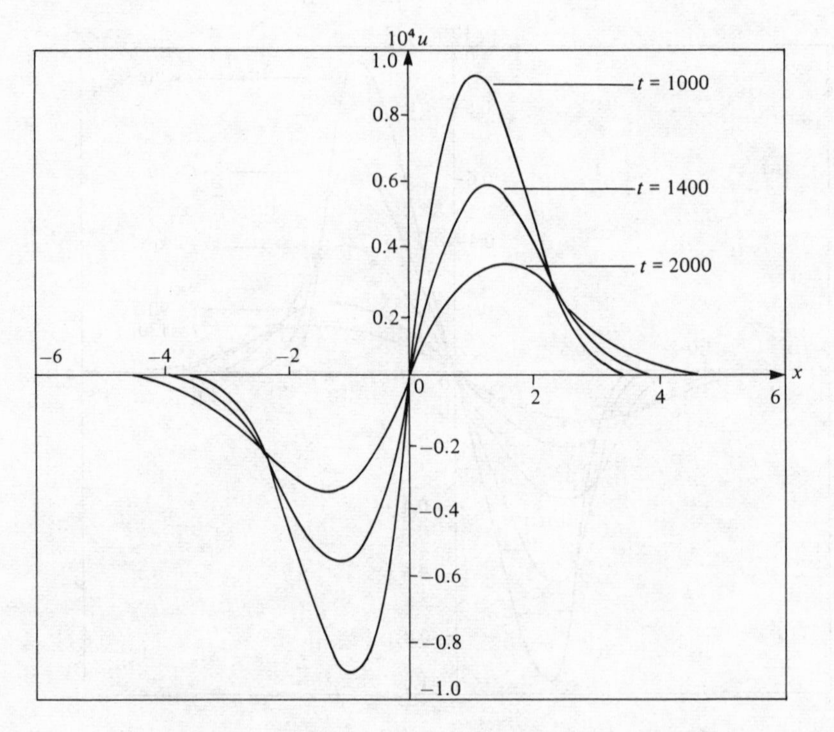

Fig. 5.12. *N* wave solution of cylindrical Burgers equation: old-age régime.

Table 5.13. *Reynolds numbers for cylindrical Burgers equation at different times: numerical, and analytic according to eqs.* (5.44) *and* (5.61) *with* $t_0 = 692.08$, $\tilde{c}^{1/2} = -4.716$ *and* $b_3^{-1/2} = 26.44$. *Numerical and analytic* (*eq.* (5.42)) *values of* $u_x(0, t)$ *are also given.*

	R			$u_x(0,t)$	
t	Numerical	eq. (5.44)	eq. (5.61)	Numerical	Analytic
5.0	6.52471	6.52365	5.61662	0.10039	0.10000
10.0	4.46658	4.46647	3.84512	0.05014	0.05000
20.0	3.01153	3.01183	2.59951	0.02502	0.02500
50.0	1.72225	1.72107	1.50794	0.00976	0.01000
100.0	1.08011	1.07054	0.97085	0.00447	0.00500
150.0	0.80369	0.78234	0.73931	0.00269	0.00333
200.0	0.64448	0.61054	0.60448	0.00182	0.00250
250.0	0.53965	0.49329	0.51440	0.00132	0.00200
350.0	0.40882	0.33949	0.39940	0.00080	0.00143
450.0	0.32989	0.24036	0.32781	0.00053	0.00111
550.0	0.27683	0.16968	0.27831	0.00038	0.00091
650.0	0.23854	0.11602	0.24178	0.00029	0.00077
750.0	0.20969	0.07348	0.21357	0.00022	0.00067
850.0	0.18711	0.03869	0.19106	0.00018	0.00059
950.0	0.16885	0.00955	0.17264	0.00015	0.00053
1050.0	0.15374	−0.01582	0.15726	0.00012	0.00048
1150.0	0.14132	−0.03687	0.14420	0.00010	0.00043
1250.0	0.13072	−0.05579	0.13297	0.00009	0.00040
1350.0	0.12160	−0.07255	0.12320	0.00008	0.00037
1450.0	0.11367	−0.08756	0.11462	0.00007	0.00034
1550.0	0.10671	−0.10108	0.10702	0.00006	0.00032
1650.0	0.10055	−0.11336	0.10024	0.00005	0.00030
1700.0	0.09773	−0.11909	0.09712	0.00005	0.00029
1800.01	0.09254	−0.12982	0.09133	0.00005	0.00028
1900.0	0.08787	−0.13969	0.08610	0.00004	0.00026
2000.01	0.08364	−0.14882	0.08133	0.00004	0.00025

$b_3^{-1/2} = 26.44$ and $\tilde{c}^{1/2} = -4.716$. At a later stage of the evolution of the N wave, the linear (old-age) solution

$$u = c_1 x t^{-2} e^{-x^2/2\delta t} \tag{5.62}$$

holds. Here, c_1 was again found by matching with the numerical solution to be about 161.02. Table 5.14 shows the analytic and numerical values of u_{max}

Table 5.14. *'Old-age' solution for cylindrical Burgers equation: numerical, and analytic according to eq.* (5.62) *with* $c_1 = 161.02$.

	$10^4 u_{\max}$		Difference	Reynolds numbers	
t	Numerical	Old-age	(3)–(2)	Numerical	Old-age
(1)	(2)	(3)	(4)	(5)	(6)
1550	0.504	0.502	−0.002	0.1067	0.1067
1600	0.482	0.480	−0.002	0.1035	0.1034
1650	0.462	0.459	−0.003	0.1005	0.1002
1700	0.442	0.440	−0.002	0.0977	0.0972
1800	0.408	0.402	−0.006	0.0925	0.0919
1850	0.392	0.387	−0.005	0.0901	0.0890
1900	0.378	0.372	−0.006	0.0879	0.0871
1950	0.364	0.358	−0.006	0.0857	0.0848
2000	0.350	0.343	−0.007	0.0836	0.0827

and Reynolds number in the old-age régime. The agreement is quite satisfactory, in spite of the long time interval over which the computation was carried out.

Spherical Burgers equation

In this case, the initial discontinuous profile is again the inviscid solution

$$u(x, t_i) = \begin{Bmatrix} x/t_i \ln t_i, & |x| < l_0, \\ 0, & |x| > l_0, \end{Bmatrix} \tag{5.63}$$

where $t_i = 1.76$, $l_0 = 0.2$, and $R(t_i) = 20$. This profile evolves through embryonic shock stage to one with a Taylor shock as shown in fig. 5.13. Table 5.15 shows the shock details in the manner discussed earlier for the cylindrical case. The Crighton–Scott formula for the centre of the shock,

$$l_0 x = X = l_0 T^{1/2}\{1 - \varepsilon[\mathrm{Ei}(T/T_0) - \mathrm{Ei}(1/T_0)]\} \tag{5.64}$$

where

$$\varepsilon = \delta\, \mathrm{e}^{-1/T_0}/(\gamma + 1)U_0 l_0,$$

$$\mathrm{Ei}(x) = \int_{-\infty}^{x} t^{-1}\, \mathrm{e}^t\, \mathrm{d}t,$$

$$T = 1 + \frac{\gamma + 1}{2}(U_0 t_i / l_0) \ln (t/t_i),$$

$$T_0 = \frac{\gamma + 1}{2l_0} U_0 t_i,$$

gives an accurate description of the Taylor shock up to $t \approx 16$ with an error less than 5%. At about this time the shock width is $O(l_0)$. Figs. 5.13–5.15 display the typical form of the N wave when it has an embryo/Taylor shock structure, a thick shock, and is in its old age, respectively. The Reynolds number as given by the numerical solution, eq. (5.44), and by

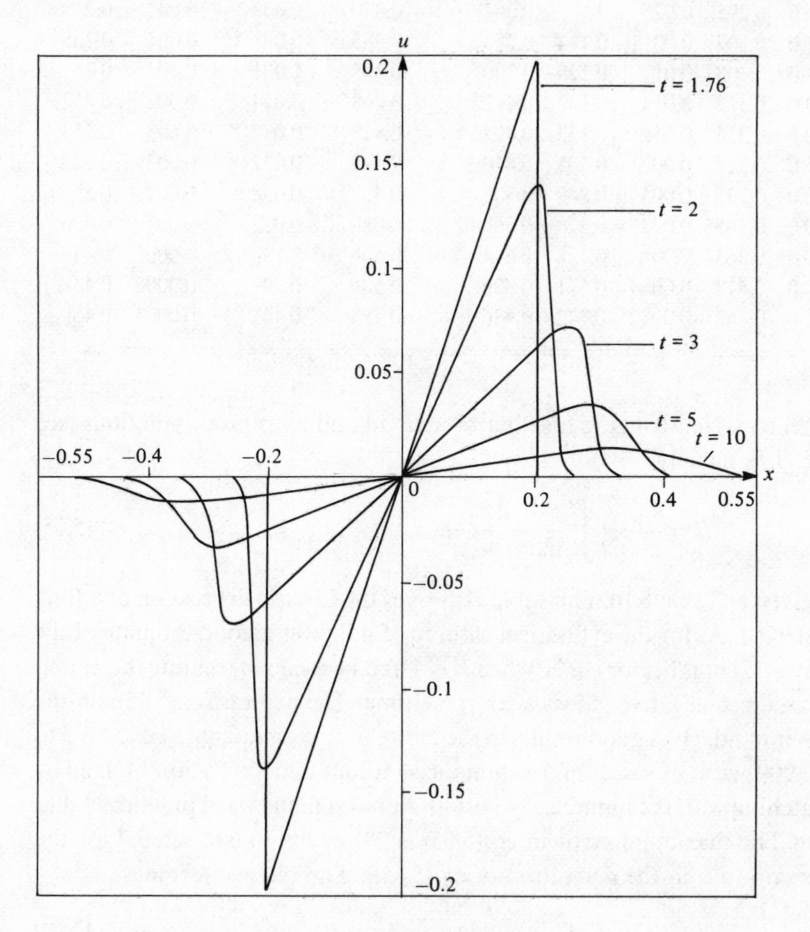

Fig. 5.13. N wave solution of spherical Burgers equation: embryo shock to Taylor shock.

Table 5.15. *Shock details for spherical Burgers equation:* $T_0 = 2.12$, $\varepsilon = 0.0065$ (*see eq.* (5.64)); *abbreviations are explained in Table* 5.10.

t (1)	x_{max} (2)	u_{max} (3)	Shock width (4)	Shock centre Numerical (5)	Shock centre CS (6)	Difference (5)–(6) (7)	u_d (8)	u_d/u_{max} (9)
2.0	0.204	0.144	0.028	0.225	0.225	0.0	0.012	0.086
3.0	0.245	0.072	0.049	0.285	0.289	−0.004	0.013	0.185
4.0	0.270	0.046	0.068	0.321	0.326	−0.005	0.009	0.196
5.0	0.284	0.033	0.085	0.347	0.351	−0.004	0.007	0.197
6.0	0.295	0.025	0.1	0.367	0.370	−0.003	0.005	0.2
7.0	0.302	0.02	0.114	0.385	0.385	0.0	0.004	0.205
8.0	0.308	0.017	0.128	0.399	0.398	0.001	0.004	0.21
9.0	0.312	0.014	0.141	0.412	0.408	0.004	0.003	0.232
10.0	0.316	0.012	0.153	0.424	0.417	0.007	0.003	0.253
15.0	0.328	0.007	0.205	0.470	0.448	0.022	0.002	0.322
20.0	0.335	0.005	0.249	0.505	0.467	0.038	0.002	0.354
30.0	0.345	0.002	0.32	0.560	0.488	0.072	0.001	0.406
50.0	0.363	0.001	0.428	0.644	0.506	0.138	0.0005	0.44
100.0	0.412	0.0004	0.611	0.806	0.506	0.299	0.0002	0.45
150.0	0.462	0.0002	0.742	0.936	0.488	0.448	0.0001	0.458

a 'reasonable' stringing together of inviscid and asymptotic solutions (see sec. 3.7), namely

$$R = \frac{\bar{c}}{t} - \frac{1}{2t}\int_{t_i}^{t} \frac{\mathrm{d}y}{\ln y + a_0 y^{3/2}}, \tag{5.65}$$

is given in Table 5.16. This table also gives the inviscid expression of $u_x(0, t) = 1/t \ln t$. As for the cylindrical case, eq. (5.44) gives a good estimate of the Reynolds number for $t \approx 30$ when $R \sim 1$ and becomes inaccurate thereafter. It assumes negative values after some time. The expression (5.65), on the other hand, gives good results in the entire time régime from about $t = 5$ to $t = 900$, with the values of constants, $a_0 = 0.0028$ and $\bar{c} = 35.50$, obtained by matching with the numerical solution. At $t = 900$, the wave practically dies out. The maximum error in eq. (5.65) is 2% at $t \approx 50$ (see sec. 3.7 for the assumptions in the derivation of eq. (5.65)). The old-age formula

$$u = c_2 x t^{-5/2} \mathrm{e}^{-x^2/2\delta t}, \tag{5.66}$$

with $c_2 = 263.92$ obtained from matching with the numerical solution,

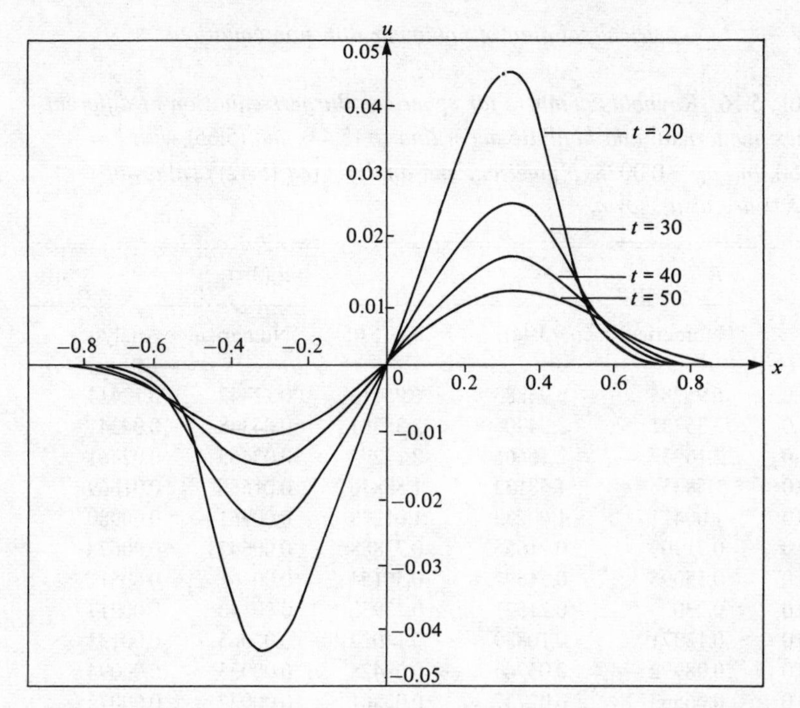

Fig. 5.14. *N* wave solution of spherical Burgers equation: thick shock régime.

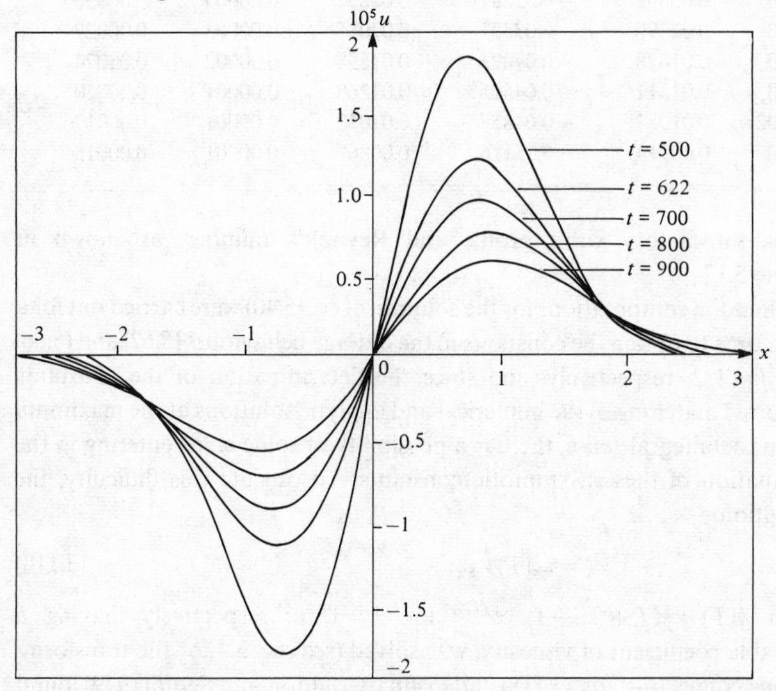

Fig. 5.15. *N* wave solution of spherical Burgers equation: old age régime.

Table 5.16. *Reynolds numbers for spherical Burgers equation at different times: numerical, and analytic according to* (5.44) *and* (5.65) *with* $\bar{c} = 35.50$ *and* $a_0 = 0.0028$. *Numerical and analytic (eq.* (5.42)) *values of* $u_x(0, t)$ *are also shown.*

	R			$u_x(0,t)$	
t	Numerical	eq. (5.44)	eq. (5.65)	Numerical	Analytic
5.0	6.95788	6.94885	6.95113	0.12442	0.12414
10.0	3.35331	3.34906	3.35361	0.04348	0.04341
15.0	2.16937	2.16666	2.17273	0.02463	0.02461
20.0	1.58335	1.58133	1.58876	0.01666	0.01669
30.0	1.00411	1.00233	1.01228	0.00961	0.00980
40.0	0.71909	0.71656	0.72888	0.00642	0.00678
50.0	0.55095	0.54697	0.56154	0.00461	0.00511
100.0	0.23003	0.21521	0.23932	0.00143	0.00217
150.0	0.13371	0.10889	0.14022	0.00065	0.00133
200.0	0.08992	0.05744	0.09428	0.00035	0.00094
250.0	0.06563	0.02747	0.06861	0.00022	0.00072
350.0	0.04047	−0.00546	0.04179	0.00010	0.00049
450.0	0.02808	−0.02281	0.02851	0.00005	0.00036
550.0	0.02090	−0.03331	0.02085	0.00003	0.00029
650.0	0.01628	−0.04023	0.01598	0.00002	0.00024
750.0	0.01311	−0.04505	0.01268	0.00001	0.00020
850.02	0.01083	−0.04857	0.01032	0.00001	0.00017
948.0	0.00914	−0.05116	0.00860	0.00001	0.00015

gives satisfactory wave profile and Reynolds number, as shown in Table 5.17.

Since the computations for the solution of eq. (5.40) were carried out for a long time to obtain the constants in the old-age behaviours (5.62) and (5.66) for $J = 1, 2$, respectively, and since the determination of the constants required matching of the numerical and analytic solutions at the maximum or in the integral sense, there is a possibility of some error entering in the evaluation of these asymptotic constants. To obviate this difficulty, the equation

$$\bar{V}_T + \bar{V}\bar{V}_X = \varepsilon g(T)\bar{V}_{XX}, \qquad (3.118)$$

with $g(T) = \frac{1}{2}(T + T_0 - 1)$, e^{T/T_0} for $J = 1$, 2, respectively, having a variable coefficient of viscosity, was solved (see sec. 3.4 for the transformations connecting eqs. (3.118) and (5.40)). Crighton and Scott (1979) found

Table 5.17. *'Old-age' solution for spherical Burgers equation: numerical, and analytic according to eq.* (5.66) *with* $c_2 = 263.92$.

	$10^4 u_{max}$			Reynolds number	
t	Numerical	Old age	Difference (3)–(2)	Numerical	Old age
(1)	(2)	(3)	(4)	(5)	(6)
550	0.167	0.167	0.0	0.0204	0.0209
622	0.132	0.131	−0.001	0.0174	0.0174
650	0.121	0.120	−0.001	0.0163	0.0163
700	0.105	0.103	−0.002	0.0146	0.0146
750	0.092	0.090	−0.002	0.0131	0.0131
800	0.080	0.079	−0.001	0.0119	0.0119
850	0.072	0.070	−0.002	0.0108	0.0109
900	0.064	0.062	−0.002	0.0099	0.0100
948	0.058	0.056	−0.002	0.0091	0.0092

that the old-age solution for eq. (3.118) is given by

$$\bar{V} = C_c \frac{X}{\varepsilon^2 T^3} e^{-X^2/\varepsilon T^2}, \quad J = 1, \tag{5.62a}$$

$$\bar{V} \sim \frac{1}{6\sqrt{\pi}} T_2^{1/2} \frac{X}{(\varepsilon T_0 e^{T/T_0})^{3/2}} \exp\left(-\frac{X^2}{4\varepsilon T_0 e^{T/T_0}}\right),$$

or

$$\bar{V} = C_s \frac{1}{6\sqrt{\pi}} T_2^{1/2} \frac{X}{(\varepsilon T_0 e^{T/T_0})^{3/2}} \exp\left(-\frac{X^2}{4\varepsilon T_0 e^{T/T_0}}\right),$$

$$J = 2, \tag{5.66a}$$

(see sec. 3.4 for definition of parameters). The following initial data were assumed:

$$J = 1, \quad T_0 = 1.2, \quad \varepsilon = 0.017; \quad J = 2, \quad T_0 = 2.12, \quad \varepsilon = 0.0065;$$
$$\Delta x = 0.005, \quad \Delta T = 0.0001;$$

number of mesh points in the initial N wave = 400, total mesh points = 512.

The discontinuous profile (3.119) was tackled initially by PS method with a very small time mesh size, but as soon as the profile smoothed, IMP was resorted to. The mesh sizes in the latter case were increased to $\Delta x = 0.01$ and $\Delta T = 0.005$ as the N wave broadened and assumed relatively small gradients everywhere. Unlike the solution for eq. (5.40), the solution for eq. (3.118) required the inversion of a very large order (1000–2500) matrix in the thick shock régime. This cost considerable computer time. The details of the old-age solution were found to be as follows:

J	Approximate old-age onset time T	Asymptotic constants	Maximum amplitude, $\bar{V}_{\max}$
1	$99(t=3417)$	$C_c=0.34$	$0(7\times10^{-3})$
2	$11(t=4160)$	$C_s=0.67$	$0(3\times10^{-3})$

The onset times here are small in comparison with those for eq. (5.40); the latter have been shown alongside in round brackets. Both eq. (5.62a) and eq. (5.66a) give excellent descriptions of the old-age evolution of the wave, which agrees with the numerical solution very closely – almost to six decimal places. It may be noted that the constant T_2 in (5.66a) heralds the onset of old age and is a root of the equation $\varepsilon T_2^{-1} e^{T_2/T_0}=1$ (see eq. (3.47) of Crighton and Scott (1979)). It is found to be approximately 9.2. At this time, the shock width $\varepsilon^{1/2} e^{(1/2)T_2/T_0}$ is almost equal to the scale $T_2^{1/2}$ of the main N wave, its value being 3.0331 at $T_2=9.2$ (see again eq. (3.47) of Crighton and Scott). The old age sets in at $T_2\approx 11$, showing that the preceding (non-Taylor) evolutionary shock régime persists for a rather short time.

5.7 Generalised Burgers equation with damping

We now consider the GBE

$$u_t + u^\beta u_x + \lambda u^\alpha = \frac{\delta}{2} u_{xx}, \tag{5.67}$$

where β and α are real constants; the additional term λu^α represents damping, positive or negative. A special case of this equation with $\beta=1$, $\alpha=1$ was studied by Lardner and Arya (1980) (see sec. 3.2) using matched asymptotic expansions. This particular case, and its extensions, describe motions of a continuous medium for which the stress–strain relation contains a large linear term proportional to the strain, a small term which is quadratic (and/or cubic) in the strain and a small dissipative term proportional to the strain-rate. The λu term arises in such a system if the equation of motion includes a small viscous damping term proportional to the velocity (see also Crighton (1979)).

In what follows, we shall compare our results with a related study

(Murray, 1970a) of an inviscid form of eq. (5.67) with more general convective and damping terms, namely

$$\left.\begin{aligned} &u_t + g(u)u_x + \lambda h(u) = 0, \\ &\lambda > 0, g_u(u) > 0, h_u(u) > 0, \quad u > 0. \end{aligned}\right\} \tag{5.68}$$

This equation includes models describing stress wave propagation in a nonlinear Maxwell rod with damping, ion exchange in fixed columns and a realistic model equation which has been suggested to explain the Gunn effect in semi-conductors. In general, $\lambda > 0$, and the term $\lambda h(u) > 0$ is dissipative, but there can be interesting cases for which λ is negative (see Murray (1970b)). We summarise Murray's (large time) asymptotic results, derived under the assumptions that $h(u) = O(u^\alpha)$ and $0 < u \ll 1$. Choosing initial single hump profile (continuous or discontinuous at the front), he arrived at the following results which depend only on α and are independent of the form of $g(u)$ except for the requirement that $g_u(u) > 0$ for $u > 0$:

(i) if $0 < \alpha < 1$, the solution is unique under certain conditions and decays in a finite time and a finite distance,
(ii) if $\alpha = 1$, the solution decays in a finite distance but in an infinite time exponentially,
(iii) if $1 < \alpha \leqslant 3$, the solution decays in an infinite distance and infinite time like $O(t^{-1/[\alpha-1]})$,
(iv) if $\alpha > 3$, the solution decays like $O(t^{-1/2})$.

Murray used the characteristic solution of eq. (5.68) to arrive at these results.

We shall also make a qualitative comparison of our results for the case $\beta = 1, \alpha = 1$ of eq. (5.67) with those of Leibovich and Randall (1979) for the modified K–dV equation with a damping term,

$$\left.\begin{aligned} &u_t = \mu u u_x - u_{xxx} - \lambda u, \quad \mu > 0, \\ &u(x, 0) = f(x), \\ &u(x, t) \to 0 \text{ as } |x| \to \infty, \quad \text{for all } t < \infty. \end{aligned}\right\} \tag{5.69}$$

Here, μ and λ are constants; the damping coefficient λ may again be positive or negative.

We first study the self-similar solutions of eq. (5.67) which go to zero as $|x| \to \infty$ and discuss which of these constitute intermediate asymptotics for the solutions of the initial value problem for eq. (5.67) vanishing as $|x| \to \infty$. (The parameter α will play a crucial role here as in the study of Murray 1970a). For this purpose, we shall solve the initial value problem for eq. (5.67), using the pseudo-spectral finite difference approach when the initial profile is discontinuous at the front and the implicit finite difference scheme when it is smooth (or it has evolved to become smooth). The present study will further enhance our understanding of self-similar phenomena for nonlinear diffusion equations. The present account is due to Sachdev, Nair and Tikekar (1986). A preliminary study of this problem was carried out by Sachdev (1979).

As we have discussed in detail in sec. 2.6, the Burgers equation (with $\beta = 1, \lambda = 0$ in eq. (5.67)) has the single hump solution

$$u = \left(\frac{\delta}{t}\right)^{1/2} \frac{1}{\sqrt{2\pi/(e^R - 1)}\cdot e^{\xi^2} + \sqrt{(\pi/2)}\cdot e^{\xi^2}\operatorname{erfc}\xi} \tag{5.70}$$

$$\equiv \left(\frac{\delta}{t}\right)^{1/2} \frac{1}{H_B(\xi)}, \tag{5.71}$$

where

$$\xi = \frac{x}{(2\delta t)^{1/2}}. \tag{5.72}$$

Whitham (1974) has shown that the solution (5.70) arises from the singular initial condition $u(x, 0) = u_0 + A\delta(x)$ where u_0 and A are constants. A change of variables, $u = u_0 + \tilde{u}, x = u_0 t + \tilde{x}$, leaves the Burgers equation invariant and changes the initial condition to $A\delta(\tilde{x})$ (self-similar solutions are typically born of this kind of singular initial conditions). We, therefore, assume that the initial condition that gives rise to the solution (5.70) is $A\delta(x)$. The solution (5.70) describes a single pulse (a hump) whose length increases with time but whose Reynolds number $R = (1/\delta) \int_{-\infty}^{\infty} u\,dx$ is constant for all time (see sec. 2.6).

The solution (5.70) suggests that we seek for eq. (5.67) a self-similar solution of the form

$$u = t^{a_1} f([2\delta]^{-1/2} x t^{b_1}) \tag{5.73}$$

where a_1 and b_1 are real constants. Substitution of (5.73) into eq. (5.67) immediately gives $a_1 = 1/(1 - \alpha)$, $b_1 = -\frac{1}{2}$ so that the former becomes

$$u = t^{1/(1-\alpha)} f(\xi), \tag{5.74}$$

provided

$$\beta = \frac{\alpha - 1}{2}. \tag{5.75}$$

Eq. (5.67) then reduces to a nonlinear ordinary differential equation in f:

$$f'' + 2\xi f' - \frac{4}{1-\alpha} f - 4(2\delta)^{-1/2} f^{(\alpha-1)/2} f' - 4\lambda f^{\alpha} = 0. \tag{5.76}$$

This equation, in general, involves fractional powers of f. The form (5.74) shows that the similarity solution decays explicitly with time if $\alpha > 1$ and grows if $\alpha < 1$. In order to remove the fractional powers in eq. (5.76) and generalise the 'reciprocal' function H_B defined by eq. (5.71), we introduce the transformation

$$H = \delta^{1/2} f^{(1-\alpha)/2} \tag{5.77}$$

so that eq. (5.76) changes into

$$HH'' - 2(1+\alpha_1)H'^2 + 2\xi HH' - 2H^2 - 2\sqrt{2}\cdot H' - 2\lambda_1 = 0 \tag{5.78}$$

where

$$\alpha_1 = \frac{3-\alpha}{2(\alpha-1)}, \quad \lambda_1 = \lambda\delta(1-\alpha). \tag{5.79}$$

For the Burgers equation ($\alpha = 3, \lambda = 0$), eq. (5.78) reduces to

$$HH'' - 2H'^2 + 2\xi HH' - 2H^2 - 2\sqrt{2}\cdot H' = 0 \tag{5.80}$$

whose solution $H_B(\xi)$ given by (5.71) is expressible in terms of exponential function and erfc. In fact, we conjecture that the GBEs may be characterised by ordinary differential equations of the form

$$yy'' + ay'^2 + \tilde{f}(\xi)yy' + \tilde{g}(\xi)y^2 + by' + c = 0, \tag{5.81}$$

where $\tilde{f}(\xi)$ and $\tilde{g}(\xi)$ are sufficiently smooth arbitrary functions and a, b and c are real constants, in the same manner as the K–dV type equations are by Painlevé transcendents. The latter are governed by six of fifty nonlinear ordinary differential equations of second order whose only movable singularities are poles (see Hille (1969)). Eq. (5.78) is a special case of eq. (5.81) with $a = -2(1+\alpha_1) = (1+\alpha)/(1-\alpha), \tilde{f}(\xi) = 2\xi, \tilde{g}(\xi) = -2, b = -2\sqrt{2}, c = -2\lambda_1$. Actually, eq. (5.81) generalises the equation

$$yy'' + ay'^2 + \tilde{f}(\xi)yy' + \tilde{g}(\xi)y^2 = 0, \tag{5.82}$$

studied by Euler and Painlevé (see Kamke (1943, eq. 6.129, p. 574)) and is readily linearised by writing $y = v^{1/(a+1)}$ so that it becomes

$$v'' + \tilde{f}v' + (a+1)\tilde{g}v = 0. \tag{5.83}$$

Eq. (5.81) has two additional terms, by' and c, besides those in eq. (5.82). We refer to the solutions of eq. (5.81) as Euler–Painlevé transcendents. Eq. (5.78) corresponding to the GBE (5.67) differs from eq. (5.80) for the Burgers equation in two 'simple' ways: the numerical coefficient of H'^2 is now $-2(1+\alpha_1)$ instead of -2 and an additional constant term λ_1 is present. But, in the context of nonlinear differential equations, these 'simple' changes make a drastic difference. Indeed, eq. (5.78) does not, in general, seem to be integrable in terms of known functions. This equation is included in eq. (5.81) as a special case – hence the name Euler–Painelevé transcendents for the solutions of the former. Actually, it may be easily verified that the solutions of the non-planar GBE

$$u_t + u^\beta u_x + \frac{ju}{2t} = \frac{\delta}{2} u_{xx}, \quad j = 1, 2, \tag{5.84}$$

has a similarity form $u = t^{-1/2\beta} F(\xi)$. The function

$$H = \delta^{1/2}[F(\xi)]^{-\beta} \tag{5.85}$$

is again found to satisfy a special case of eq. (5.81), namely

$$HH'' - \frac{\beta+1}{\beta} H'^2 + 2\xi HH' - 2(1 - j\beta)H^2 - 2\sqrt{2}\cdot H' = 0. \tag{5.86}$$

Indeed, this equation differs from eq. (5.80) for the Burgers equation merely in some of the constant coefficients being different; yet it does not seem to be generally integrable. We emphasise that the class of equations (5.81) is much nicer than the Painlevé equations and displays smoother structure in the physically realistic cases.

We now return to the self-similar solutions of eq. (5.67), as represented by eqs. (5.74)–(5.76). The linearised form of eq. (5.76) is

$$f'' + 2\xi f' - \frac{4}{1-\alpha} f = 0. \tag{5.87}$$

Its solution is

$$f = A\mathrm{e}^{-\xi^2} H_\nu(\xi) \sim A\,\mathrm{e}^{-\xi^2}(2\xi)^{2\alpha_1} \quad \text{as } \xi \to +\infty, \tag{5.88}$$

where

$$\nu = 2\alpha_1 = \frac{3-\alpha}{\alpha-1}.$$

Here, H_ν denotes the Hermite function and A is the amplitude parameter.

The solution of (5.87) has the asymptotic form

$$f \sim O(\xi^{-2\alpha_1-1}) \quad \text{as } \xi \to -\infty. \tag{5.89}$$

Now, we pose a boundary value or connection problem for eq. (5.76) (see Hastings and McLeod (1980) for a related problem for second Painlevé transcendent): we seek its solution over $-\infty < \xi < \infty$, satisfying (5.88) as $\xi \to +\infty$ and tending to zero as $\xi \to -\infty$ in accordance with (5.89). It is clear from (5.88) and (5.89) that the solution decays exponentially as $\xi \to +\infty$, and algebraically as $\xi \to -\infty$ provided $2\alpha_1 + 1 > 0$, that is, $\alpha > 1$. Before solving this connection problem, we note that there are two exact solutions of eq. (5.76). The first is a constant solution

$$f = [\lambda(\alpha - 1)]^{1/(1-\alpha)} \equiv f_{\max}, \quad \text{say.} \tag{5.90}$$

It is easy to check that $f_{\max}$ is also the maximum value of f that the maxima of the single hump solutions can attain. This follows from eq. (5.76) by noting that, at the maximum, $f' = 0, f'' < 0$ etc.

The second exact solution of eq. (5.76) is

$$f = \begin{cases} (A_+\xi/\delta^{1/2})^{2/(1-\alpha)}, & \xi > 0, \\ (A_-|\xi|/\delta^{1/2})^{2/(1-\alpha)}, & \xi < 0, \end{cases} \tag{5.91}$$

where

$$\left.\begin{aligned} A_+ &= \frac{\sqrt{2}\cdot(\alpha - 1)}{\alpha + 1}\{[1 + \lambda\delta(1+\alpha)]^{1/2} + 1\}, \\ A_- &= \frac{\sqrt{2}\cdot(\alpha - 1)}{\alpha + 1}\{[1 + \lambda\delta(1+\alpha)]^{1/2} - 1\}. \end{aligned}\right\} \tag{5.92}$$

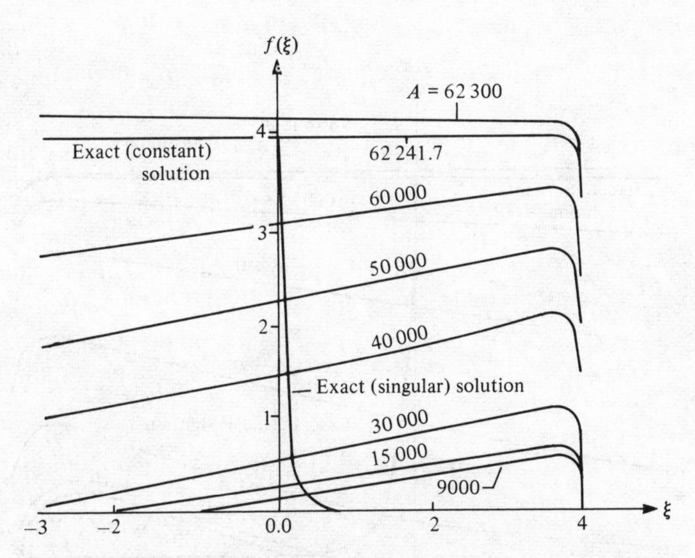

Fig. 5.16. Solution of eq. (5.76) for various values of A for $\alpha = 1.5$, $\lambda = 1$. Singular solution (5.91) for $\xi > 0$ and the constant solution (5.90) are also shown.

The corresponding solution for eq. (5.78) is

$$H(\xi) = \begin{Bmatrix} A_+\xi, & \xi > 0, \\ -A_-\xi, & \xi < 0. \end{Bmatrix} \tag{5.93}$$

The solution (5.91) is singular at $\xi = 0$ for $\alpha > 1$ and can be embedded in a two-parameter family of solutions about $\xi = \infty$ (see Hille (1970) for kindred embedding theorems for the singular solution of the Thomas–Fermi equation; see also Bender and Orszag (1978) for a discussion of the numerical solution of a boundary value problem for the Thomas–Fermi equation).

We integrated eq. (5.76) numerically, starting from $\xi \sim 4$ with initial conditions from (5.88). The amplitude parameter A was chosen rather arbitrarily. The integration was carried out in the direction of decreasing ξ until f was essentially zero. Figs. 5.16–5.19 show the solution for a set of values of $\alpha = 1.5, 2.0, 2.5$ and 3.0 (in the similarity range of α), and the corresponding values of $\beta = (\alpha - 1)/2 = 0.25, 0.5, 0.75$ and 1.0, respectively. For each such pair (α, β) there is a value of $A = A_{\max}$ for which the integral curve does not decrease to zero as $\xi \to -\infty$ but, instead, continues to rise to asymptote to a constant. This constant is just the solution $f_{\max}$ given by eq. (5.90). It is interesting to note that $f_{\max}^{\alpha-1}$ is equal to B, a constant which occurs prominently in the discussion of the asymptotic solution in Murray's

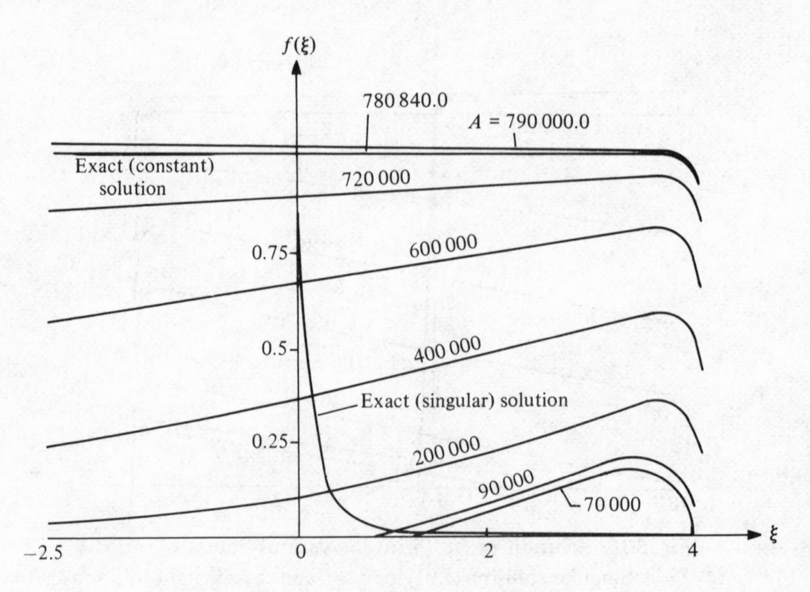

Fig. 5.17. Same as in Fig. 5.16, $\alpha = 2.0, \lambda = 1$.

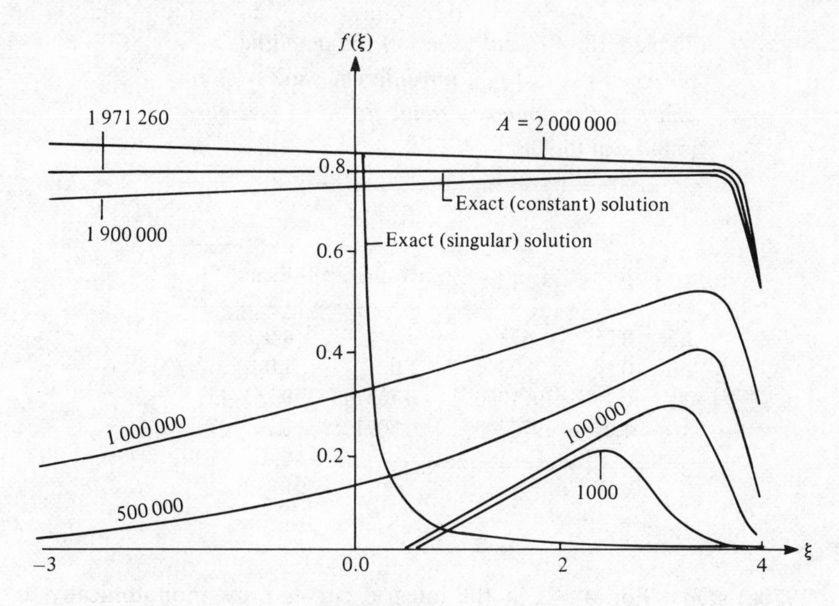

Fig. 5.18. Same as in Fig. 5.16, $\alpha = 2.5, \lambda = 1$.

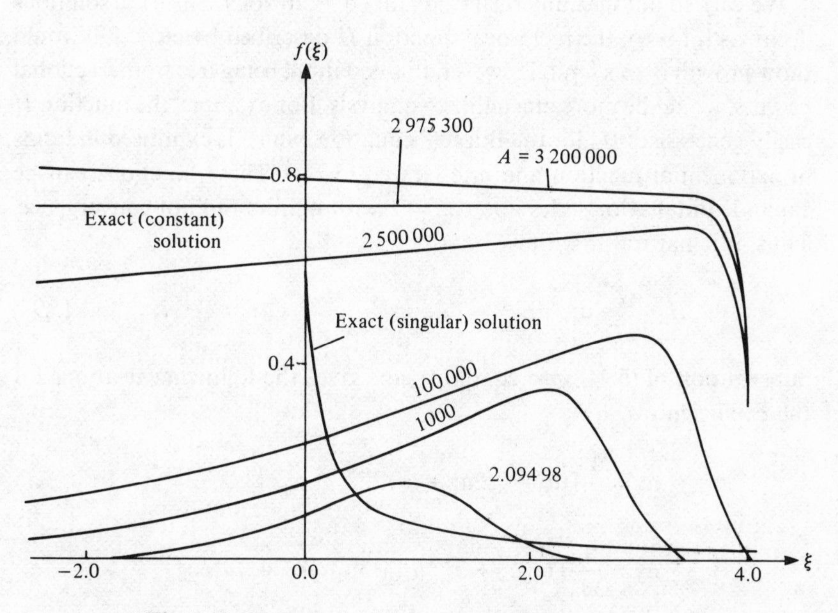

Fig. 5.19. Same as in Fig. 5.16, $\alpha = 3.0, \lambda = 1$.

Table 5.18. *Critical values of the amplitude parameter A and f_{max} for different choices of α and β in the similarity range for $\lambda = 1$ (see eqs. (5.88) and (5.90)).*

			f_{max}	
α	β	A_{max}	Numerical	Exact
1.5	0.25	62242	4.0	4.0
2.0	0.5	780840	1.0	1.0
2.5	0.75	1971260	0.763143	0.763143
3.0	1.0	2975300	0.707128	0.707107

(1970a) study. For $A > A_{max}$ the integral curves grow monotonically to infinity as $\xi \to -\infty$ (cf. Bender and Orszag (1978, fig. 4.10) for the Thomas–Fermi equation). Table 5.18 gives the values of A_{max} for different (α, β) pairs. Figs. 5.16–5.19 also include the singular solution (5.91)–(5.92). The present study is similar to Miles' (1978) for the second Painlevé transcendent.

We carried out the numerical study for eq. (5.76) for f. Since the solutions decay as $|\xi| \to \infty$, the reciprocal function H described by eq. (5.78) would show growth in this limit. However, this equation, being free from fractional powers, would be more amenable to analysis. For example, the function H easily generalises H_B for the Burgers equation, which is expressed in terms of exponential function and erfc (see eq. (5.71)). These functions can be expanded in Taylor series about $\xi = 0$, with infinite radii of convergence. Thus, it is natural to write

$$H = \sum_{n=0}^{\infty} a_n \xi^n. \tag{5.94}$$

Substitution of (5.94) into eq. (5.78) etc., gives the following relations for the coefficients a_i:

$$a_2 = \frac{1}{a_0}[(a_0^2 + \sqrt{2}a_1 + a_1^2) + \lambda_1 + \alpha_1 a_1^2],$$

$$a_3 = \frac{1}{3a_0}[(a_0 a_1 + 2^{3/2} a_2 + 3a_1 a_2) + 4\alpha_1 a_1 a_2],$$

$$a_{k+2} = \frac{2a_k}{(k+1)(k+2)} + \frac{2a_{k+1}}{(k+2)a_0}(\sqrt{2} + \alpha_1 a_1 + a_1)$$
$$+ \frac{2}{(k+1)(k+2)a_0} \sum_{i=1}^{k} \left[-\frac{(k+1-i)(k+2-i)}{2} a_i a_{k+2-i} \right.$$
$$+ (1+\alpha_1)(i+1)(k+1-i)a_{i+1}a_{k+1-i}$$
$$\left. + a_i a_{k-i} - (k+1-i)a_{i-1}a_{k+1-i} \right], \quad k = 1, 2, \ldots. \quad (5.95)$$

Thus, we have a two-parameter a_0, a_1 family of series solutions. The convergence of this series by direct computation seems difficult to establish. For the Burgers equation with $\lambda = 0$, $\alpha = 3$, the function H_B defined by eq. (5.71) follows from eq. (5.94) if we choose $a_1 = -\sqrt{2}$. (Here the expansion

$$\operatorname{erfc} z = 1 - \operatorname{erf} z = 1 - \frac{2}{\sqrt{\pi}} e^{-z^2} \sum_{n=0}^{\infty} \frac{2^n}{1 \cdot 3 \cdots (2n+1)} z^{2n+1}$$

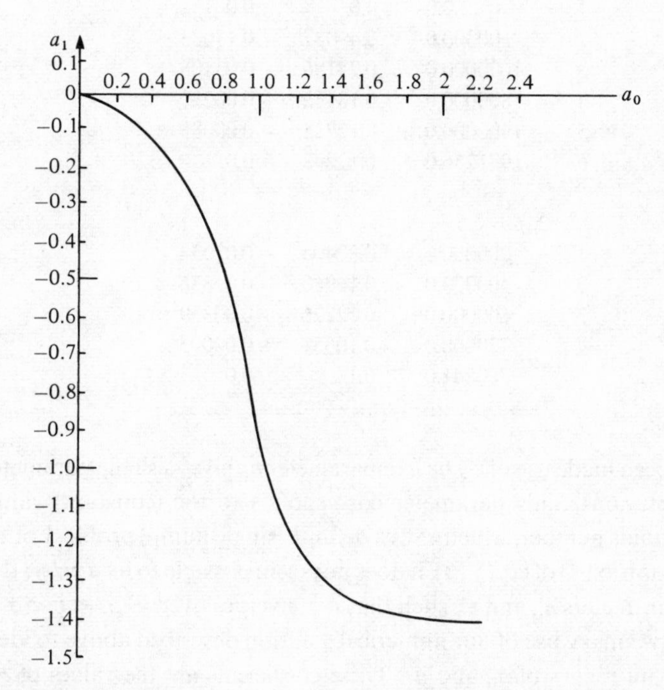

Fig. 5.20. a_1 versus a_0 for $0 \leqslant A \leqslant A_{max}$ (see eq. (5.94)) for $\alpha = 3$, $\beta = 1$ and $\lambda = 1$.

Table 5.19. *Coefficients a_0 and a_1 in the series (5.94) for the permissible (similarity) range of the amplitude parameter A corresponding to different values of α and β and for $\lambda = 1$,* $\delta = 0.01$.

No.	A	a_0	a_1
$\alpha = 3, \beta = 1$			
1	0.0	0.0	0.0
2	2.095	1.2182	−1.1953
3	3.25	1.4233	−1.3305
4	1000.0	0.70023	−0.36571
5	100000.0	0.44529	−0.13077
6	2500000.0	0.16709	−0.00555
7	2953000.0	0.141414	0.0
$\alpha = 2.5, \beta = 0.75$			
1	0.0	0.0	0.0
2	500000.0	0.49037	−0.17083
3	1000000.0	0.24186	−0.03108
4	1800000.0	0.13462	−0.0022
5	1900000.0	0.12728	−0.00085
6	1971256.0	0.12248	0.0
$\alpha = 2, \beta = 0.5$			
1	200000.0	0.32406	−0.07034
2	400000.0	0.16960	−0.01336
3	600000.0	0.12276	−0.00359
4	720000.0	0.10651	−0.00095
5	780841.0	0.1	0.0

has been made use of). The free parameter a_0 gives a single parameter family of solutions. This parameter corresponds to the (constant) value of the Reynolds number, which fixes a definite (single hump) profile. For the series solution (5.94) of eq. (5.78), it does not seem possible to fix *a priori* the ranges of parameters a_0 and a_1 such that it converges over $-\infty < \xi < \infty$. To this end, we make use of our numerical solution described above to identify the relevant ranges of a_0 and a_1. These coefficients are the values of H and its derivative at $\xi = 0$. The function H is related to f by eq. (5.77). Therefore, for each value of A, $0 < A < A_{max}$, we can find from the numerical solution of f

Table 5.20. *Comparison of series solution* (5.94) *and numerical solution of eq.* (5.76) *for* $\alpha = 3$, $\beta = 1, \lambda = 1$ *and* $\delta = 0.01$. *The* a_i *in eq.* (5.94) *from* $i = 0$ *to* $i = 13$ *are* 1.2182, – 1.1957, 0.98736, – 0.60357, 0.34600, – 0.14134, 0.05471, – 0.00713, – 0.00259, 0.00537, – 0.00297, 0.00155, – 0.00044, 0.00007.

	Series solution		Numerical solution
	$H(\xi)$	$f(\xi)$	$f(\xi)$
– 3.0	36.77590	0.0027192	0.0027192
– 2.5	29.15179	0.0034303	0.0034303
– 2.0	19.82509	0.0050441	0.0050441
– 1.5	10.26901	0.0097380	0.0097380
– 1.0	4.541242	0.0220204	0.0220204
– 0.5	2.165305	0.0461829	0.0461829
0.0	1.218223	0.0820868	0.0820867
0.5	0.8097551	0.1234941	0.1234941
1.0	0.6595272	0.1516237	0.1516237
1.5	0.8547518	0.1169930	0.1169928
2.0	2.926066	0.0341755	0.0341754
2.5	24.9906	0.0040015	0.0040015
3.0	387.044	0.0002583	0.0002584
3.5	9993.162	0.0000100	0.0000100
4.0	366869.9	0.0000003	0.0000003

the value of $f(0)$ and $f'(0)$ and hence $H(0)$ and $H'(0)$. Table 5.19 contains the relevant values of a_0 and a_1 for $0 < A \leqslant A_{max}$ while fig. 5.20 shows a_1 versus a_0, for $\alpha = 3$, $\beta = 1$ and $\lambda = 1$. With a_0 and a_1 thus determined, the series (5.94) was summed up and compared with the exact numerical solution of f, using eq. (5.77) again. The series converged up to some value of ξ and then its convergence slowed down. However, analytic continuation of the series at a couple of ξ points yields excellent results over a large finite range of ξ. The discrepancy from the exact numerical solution was found to be $O(10^{-7})$ in single precision (see Table 5.20).

It is clear from the asymptotic form (5.88)–(5.89) (and has been numerically checked by us) that the solution of the connection problem for eq. (5.76) exists for all $\alpha > 1$. However, the similarity solution (5.74) of eq. (5.67) is significant only in the range $1 < \alpha \leqslant 3$ since, as we shall presently discuss, the solutions of the partial differential eq. (5.67) with 'suitable' initial conditions vanishing at $\pm\infty$ asymptote to the self-similar form only in this range of α. The reason for this is the physically unrealistic decay predicted by the similarity form (5.74) for $\alpha > 3$. Nevertheless, eq. (5.76) has single hump solutions vanishing at $\xi = \pm\infty$ for all positive λ and $\alpha > 1$. Correspondingly, eq. (5.78) has 'inverted hump' solutions growing to infinity at $\xi = \pm\infty$ for positive λ and $\alpha > 1$. The left end limit of the range $1 < \alpha \leqslant 3$, namely $\alpha = 1$, gives $\beta = 0$ according to eq. (5.75) and the solution form (5.74) does not exist in this case. However, eq. (5.67) now becomes linear, so that

$$u_t + u_x + \lambda u = \frac{\delta}{2} u_{xx}. \tag{5.96}$$

It has an exact single hump solution

$$u = \frac{A}{(2\delta t)^{1/2}} \exp\left\{-\left[\frac{1}{2\delta t}(x-t)^2 + \lambda t\right]\right\} \tag{5.97}$$

which decays exponentially with time (and distance). The form of this solution is quite different from (5.74).

When λ is negative, the asymptotic form (5.88)–(5.89) is still valid for $\alpha > 1$, but the constant solution (5.90) ceases to exist. This suggests that, in this case, there is probably no upper limit A_{max} to the amplitude that the linear solution can possess. Our numerical study of eqs. (5.76), (5.88)–(5.89) with $\lambda < 0$ and $\alpha > 1$ confirms this conclusion (see fig. 5.21). However, we shall find, as we solve the partial differential eq. (5.67) numerically, that these self-similar solutions do not constitute intermediate asymptotics. We therefore conclude that even though eq. (5.76) has a solution for the connection problem for $\alpha > 1$ and all λ, it forms an intermediate asymptotic for eq. (5.67) only for $1 < \alpha \leqslant 3$ and $\lambda > 0$.

An interesting feature that emerges from our numerical solution for $\alpha > 3$ and $\lambda \leqslant -1$ is the appearance of a shelf on the left end tail of the self-similar profile (see fig. 5.21). The solution decays in an extremely slow manner – characteristic of a shelf. Eq. (5.67) shares this feature with the modified K–dV equation (5.69) when $\lambda < 0$ (see Leibovich and Randall (1979)). However, if $\lambda \ll -1$, the shelf at the left-end tail has a tendency to disappear.

It does not seem possible to solve an initial value problem for eq. (5.67) analytically. A special case of eq. (5.67),

$$u_t + uu_x + \lambda u = \frac{\delta}{2} u_{xx}, \tag{5.98}$$

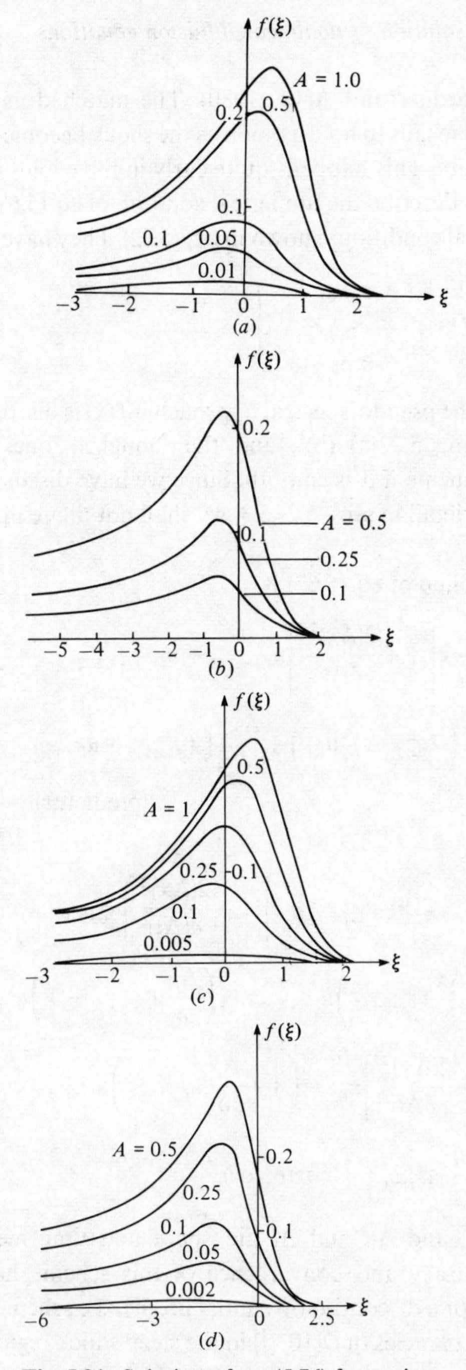

Fig. 5.21. Solution of eq. (5.76) for various values of A for $\alpha > 3$ and $\lambda < 0$. (a) $\alpha = 4$, $\lambda = -1$. (b) $\alpha = 5$, $\lambda = -1$. (c) $\alpha = 5$, $\lambda = -5$. (d) $\alpha = 4$, $\lambda = -5$.

was considered by Lardner and Arya (1980). The matched asymptotic expansion found by them fails to hold as soon as the shock becomes thick or loses its Taylor structure. This happens quite early in the evolution of the initial profile. We now describe the numerical solution of eq. (5.67) subject to any of the three initial conditions shown in fig. (5.22). They have the form

$$u(x,t_i)=\left\{\begin{array}{ll}0, & x<x_0,\\ f(x), & x_0<x<x_1,\\ 0, & x>x_1.\end{array}\right\} \tag{5.99}$$

We shall again use the pseudo-spectral approach if $f(x)$ is discontinuous at the front $x=x_1$ (fig. 5.22(a)–(b)), and the Douglas–Jones implicit predictor–corrector scheme if it is smooth. Since we have discussed these numerical schemes in detail in secs. 5.2–5.3, we shall not dilate upon them here.

The difference analogue of eq. (5.67) is

$$u_{i+1,j+1/2}-2\left[1+\frac{2(\Delta x)^2}{\delta\Delta t}\right]u_{i,j+1/2}+u_{i-1,j+1/2}$$
$$=\frac{2(\Delta x)^2}{\delta}\left(\lambda u^{\alpha}_{i,j}-\frac{2}{\Delta t}u_{i,j}\right)+\frac{\Delta x}{\delta}u^{\beta}_{i,j}(u_{i+1,j}-u_{i-1,j})$$
$$\text{(predictor)}, \tag{5.100}$$

and

$$\left(1-\frac{\Delta x}{\delta}u^{\beta}_{i,j+1/2}\right)u_{i+1,j+1}+2\left[1+\frac{2(\Delta x)^2}{\delta(\Delta t)}\right]u_{i,j+1}$$
$$+\left(1+\frac{\Delta x}{\delta}u^{\beta}_{i,j+1/2}\right)u_{i-1,j+1}=\left(\frac{\Delta x}{\delta}u^{\beta}_{i,j+1/2}-1\right)u_{i+1,j}$$
$$+2\left[1-\frac{2(\Delta x)^2}{\delta\Delta t}\right]u_{i,j}-\left(1+\frac{\Delta x}{\delta}u^{\beta}_{i,j+1/2}\right)u_{i-1,j}$$
$$+\frac{4\lambda(\Delta x)^2}{\delta}u^{\alpha}_{i,j+1/2}\quad\text{(corrector)}. \tag{5.101}$$

Here, $u_{i,j}=u(i\Delta x, j\Delta t)$, and Δx and Δt are space and time mesh sizes, respectively. The accuracy and convergence of this scheme have been discussed in sec. 5.2. For a discontinuous initial profile, the scheme (5.100)–(5.101) introduces inaccuracies of $O(10^{-2})$ in the steep shock region, which tend to vitiate subsequent computations. We therefore use a pseudo-spectral finite difference approach in the early embryonic shock régime (see sec. 5.6). The solution $u(x,t+\Delta t)$ at the new time level is obtained from the

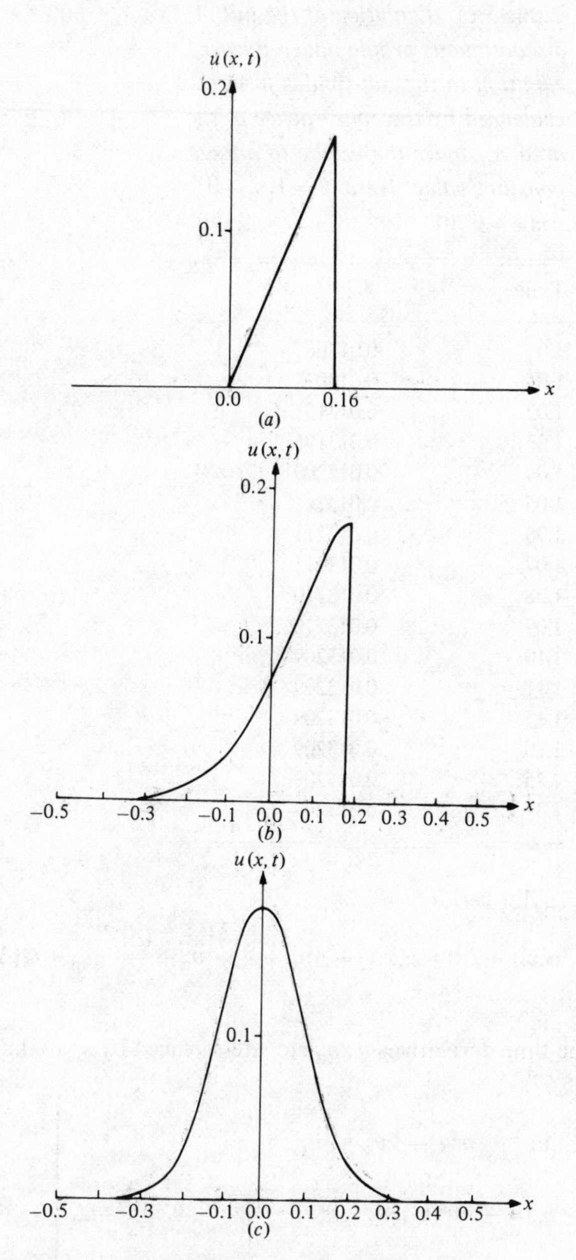

Fig. 5.22. Initial profiles for the solution of eq. (5.67).

Table 5.21. *Evolution of the initial discontinuous profile under Burgers equation to the self-similar form as evidenced by the convergence of the area A_0 under the profile to a fixed constant value. Here, $\beta = 1$, $\lambda = 0$ and $\delta = 0.001$.*

Time	A_0
1.0	0.013067
1.01	0.013078
1.02	0.013151
1.03	0.013196
1.04	0.013209
1.05	0.013211
1.06	0.013211
1.07	0.013211
1.08	0.013210
1.09	0.013210
1.10	0.013209
1.11	0.013209
1.15	0.013209
1.20	0.013209
1.25	0.013209
1.3	0.013209

truncated Taylor series

$$u(x, t + \Delta t) = u(x, t) + \Delta t u_t + \frac{(\Delta t)^2}{2!} u_{tt} + \frac{(\Delta t)^3}{3!} u_{ttt} + O(\Delta t)^4, \tag{5.102}$$

wherein the time derivatives u_t, u_{tt} etc. are replaced by spatial derivatives, using eq. (5.67):

$$\left.\begin{aligned} u_t &= -u^\beta u_x - \lambda u^\alpha + \frac{\delta}{2} u_{xx}, \\ u_{tt} &= -\beta u^{\beta-1} u_t u_x - u^\beta u_{xt} - \lambda \alpha u^{\alpha-1} u_t + \frac{\delta}{2}(u_t)_{xx}, \\ u_{ttt} &= -\beta(\beta-1)u^{\beta-2}u_t^2 u_x - 2\beta u^{\beta-1}(u_t)_x u_t \\ &\quad - \beta u^{\beta-1} u_x u_{tt} - u^\beta (u_{tt})_x \\ &\quad - \alpha(\alpha-1)\lambda u^{\alpha-2} u_t^2 - \alpha\lambda u^{\alpha-1} u_{tt} + \frac{\delta}{2}(u_{tt})_{xx}. \end{aligned}\right\} \tag{5.103}$$

Table 5.22. *Comparison of numerical (pseudo-spectral and implicit finite difference) solutions and the exact analytic solution for Burgers' equation, with smooth initial data at $t_i = 1$. Here, $\lambda = 0$, $\beta = 1$ and $\delta = 0.001$.*

	$u(x, 2)$		
x	Implicit	Pseudo-spectral	Exact
−0.10	0.000742	0.000742	0.000742
−0.08	0.001869	0.001870	0.001870
−0.06	0.003983	0.003985	0.003985
−0.04	0.007341	0.007342	0.007342
−0.02	0.012000	0.012000	0.012000
0.00	0.017843	0.017841	0.017841
0.02	0.024660	0.024657	0.024657
0.04	0.032232	0.032227	0.032227
0.06	0.040367	0.040362	0.040362
0.08	0.048918	0.048913	0.048913
0.10	0.057774	0.057767	0.057767
0.12	0.066840	0.066826	0.066826
0.14	0.075986	0.075934	0.075934
0.16	0.084648	0.084423	0.084423
0.18	0.088732	0.088036	0.088034
0.20	0.065779	0.066041	0.066041
0.22	0.017660	0.018405	0.018405
0.24	0.002151	0.002161	0.002161
0.26	0.000190	0.000180	0.000180
0.28	0.000015	0.000012	0.000012
0.30	0.000001	0.000001	0.000001

The spatial derivatives are found in the manner explained in sec. 5.6.

The (normalised) spatial interval $(0, 2\pi)$ was divided into 128 mesh points. The initial (discontinuous) non-zero profile occupied 64 points and was placed in the middle of the interval so as to allow it to grow due to diffusion as it evolves. In eq. (5.99), the function $f(x)$ was chosen to be x, and $x_0 = 0$, $x_1 = 0.2$ and $t_i = 1$. As the computation commenced, a tail $O(10^{-3})$ on either side of the non-zero part of the profile was noticed. Being spurious, it was artificially cut off. It was not found to persist later in any significant way. We first considered the Burgers equation itself corresponding to $\beta = 1$,

Table 5.23. *Comparison of the exact and numerical (IMP) solutions for the special (linear) PDE with* $\beta = 0$, $\alpha = 1$, $\lambda = 1$ *and* $\delta = 0.001$ (*see eqs.* (5.96)–(5.97)).

	$u(x, 2)$		$u(x, 4)$	
x	Numerical	Exact	Numerical	Exact
−0.11	0.00051	0.00057	0.0	0.0
0.37	0.00529	0.00546	0.0	0.0
0.85	0.02886	0.02897	0.00003	0.00003
1.33	0.08656	0.08633	0.00016	0.00017
1.81	0.14487	0.14463	0.00071	0.00072
2.29	0.13598	0.13621	0.00231	0.00233
2.77	0.07186	0.07211	0.00563	0.00562
3.25	0.02143	0.02146	0.01023	0.01019
3.73	0.00357	0.00359	0.01387	0.01383
4.21	0.00032	0.00034	0.01407	0.01409
4.69	0.00001	0.00002	0.01070	0.01075
5.17	0.0	0.0	0.00612	0.00615
5.65	0.0	0.0	0.00264	0.00264
6.13	0.0	0.0	0.00086	0.00085

$\lambda = 0$ in eq. (5.67). After a short time, the Reynolds number $R = (1/\delta)\int_{-\infty}^{\infty} u(x,t)\,\mathrm{d}x$ was found to settle down to a constant value 13.209 and the profile was found to be in excellent agreement with (5.70) with this value of R (see Table 5.21). We also checked the case with $\alpha = \beta = 1, \lambda \neq 0$ in eq. (5.67). An integration of eq. (5.67) with respect to x shows that the area under the profile $A = \int_{-\infty}^{\infty} u(x,t)\,\mathrm{d}x$ decays exponentially with time: $A = A_0 \mathrm{e}^{-\lambda t}$, where A_0 is the area under the profile after it has evolved to acquire a smooth structure. The numerical solution was found to accord to this decay law. Also smooth initial profiles were chosen – the single hump (5.70) and the exact (linear) solution (5.97) – and the implicit finite difference scheme (5.100)–(5.101) was used to continue the solution in time. The agreement of the numerical solution with these exact solutions was again found to be very good (see Tables 5.22 and 5.23). Now we summarise the numerical results. We find that, for $\lambda > 0$, $1 < \alpha \leqslant 3$, the initial profile, discontinuous or continuous at the front, soon evolves into a self-similar form discussed earlier. Fig. 5.23 shows a typical evolution of the profile to its self-similar form for the case $\alpha = 3$, $\beta = 1$, both when the initial amplitude $u_{\max}$ is less than 1 and when it is greater than 1; only $u(x, t_i)$ is required to

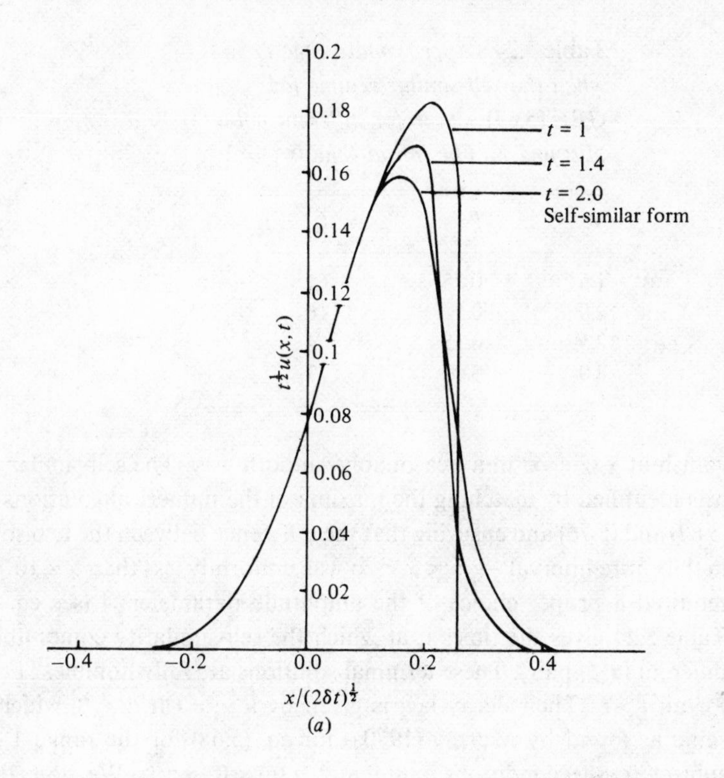

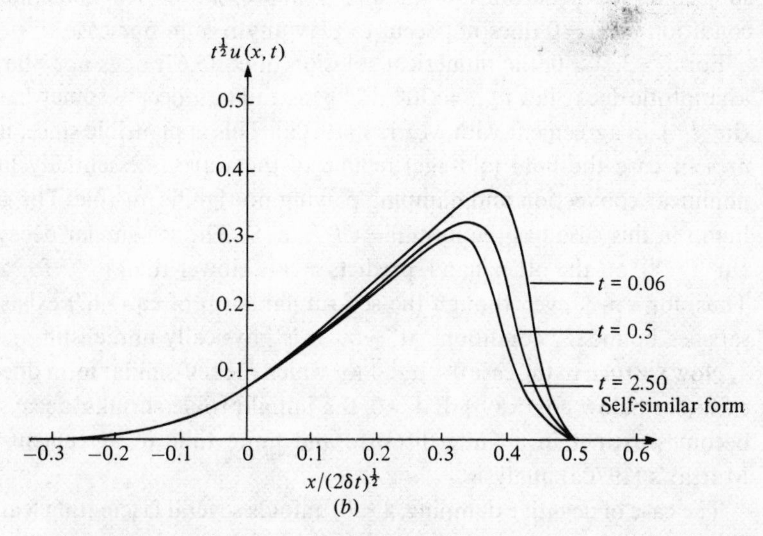

Fig. 5.23. Evolution of the solution of eq. (5.67) to self-similar form for $\alpha = 3, \beta = 1$ and $\lambda = 1$. The function $t^{1/2}u(x,t)$ is shown at various times for (a) $u_{\max}(x,t_i) < 1$, (b) $u_{\max}(x,t_i) > 1$.

Table 5.24. *Approximate time t_s when the self-similar régime for GBE (5.67) sets in for different values of α and β. The initial time is $t_i = 1$.*

α	β	t_s
1.5	0.25	61
2.0	0.50	16
2.5	0.75	4
3.0	1.00	3

vanish at $x = \pm\infty$ in a 'reasonably' smooth way. The self-similar régime was identified by matching the maxima of the numerical solutions of eqs. (5.67) and (5.76) and ensuring that the difference between the two solutions in the entire interval $-\infty < x < \infty$ was uniformly less than 5×10^{-3}. This required a proper choice of the amplitude parameter A (see eq. (5.88)). Table 5.24 gives the times t_s at which the self-similarity comes about for different (α, β) pairs. These terminal solutions are fully nonlinear and hold for all $t > t_s$. Their decay law is given by $u_{max} = O(t^{1/[1-\alpha]})$, which is the same as found by Murray (1970a) for eq. (5.68) for the range $1 < \alpha \leqslant 3$ subject to the conditions, $u \ll 1$ and $g_u(u) \neq 0$, $u > 0$. We note that the condition $g_u(u) \neq 0$ does not seem to play any role in our case.

For $\alpha > 3$, $\lambda > 0$, the numerical solution of eq. (5.67) does not obey the asymptotic decay law $u_{max} = O(t^{1/[1-\alpha]})$; instead u_{max} decays somewhat like $O(t^{-1/2})$, in agreement with Murray (1970a). This is plausible since, in the present case the final (old-age) regime of the wave is essentially linear, nonlinear convection and damping playing no significant role. The single hump in this case has the form $u = Ct^{-1/2}\mathrm{e}^{-\xi^2}$. The self-similar decay law $O(t^{1/[1-\alpha]})$, on the other hand, predicts a rate slower than $t^{-1/2}$ for $\alpha > 3$. Thus, for $\alpha > 3$, even though the self-similar form of eq. (5.67) exists and satisfies boundary conditions at $\pm\infty$, it is physically unrealistic.

Now we turn to the case $0 < \alpha < 1$ for which the self-similar form does not exist (see below eq. (5.89)). If $\lambda > 0$, the initial profile shrinks, decays and becomes extinct in a finite distance and finite time in agreement with Murray's (1970a) analysis.

The case of negative damping, $\lambda < 0$, unfolds several fascinating features. The nature of the solution again depends crucially on the parameter α. The special values $\alpha = 1$, 2 seem to demarcate distinct behaviours of the solution. We assumed λ to be -1 in all cases. For $0 \leqslant \alpha < 1$ (fig. 5.24), the

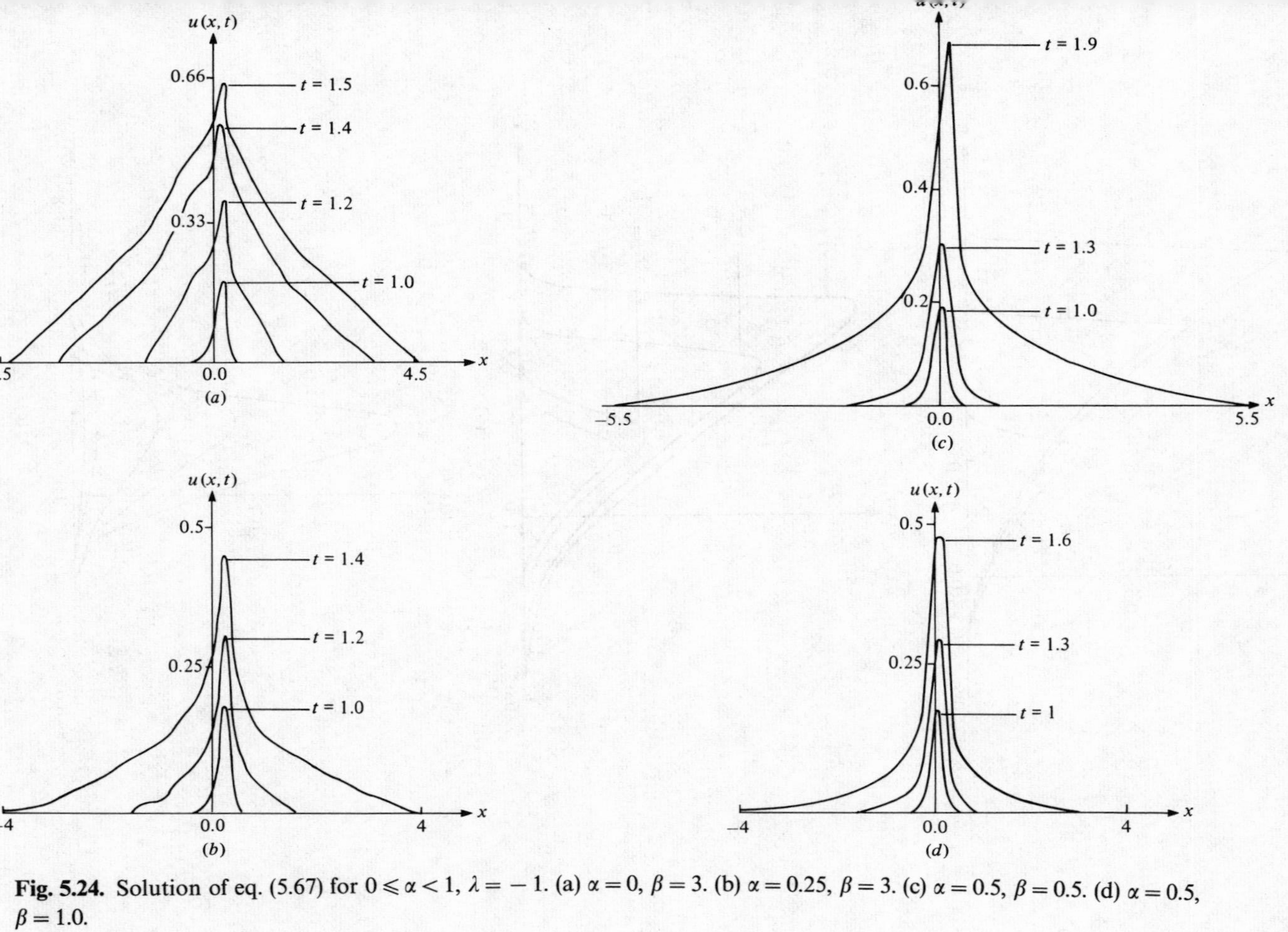

Fig. 5.24. Solution of eq. (5.67) for $0 \leqslant \alpha < 1$, $\lambda = -1$. (a) $\alpha = 0$, $\beta = 3$. (b) $\alpha = 0.25$, $\beta = 3$. (c) $\alpha = 0.5$, $\beta = 0.5$. (d) $\alpha = 0.5$, $\beta = 1.0$.

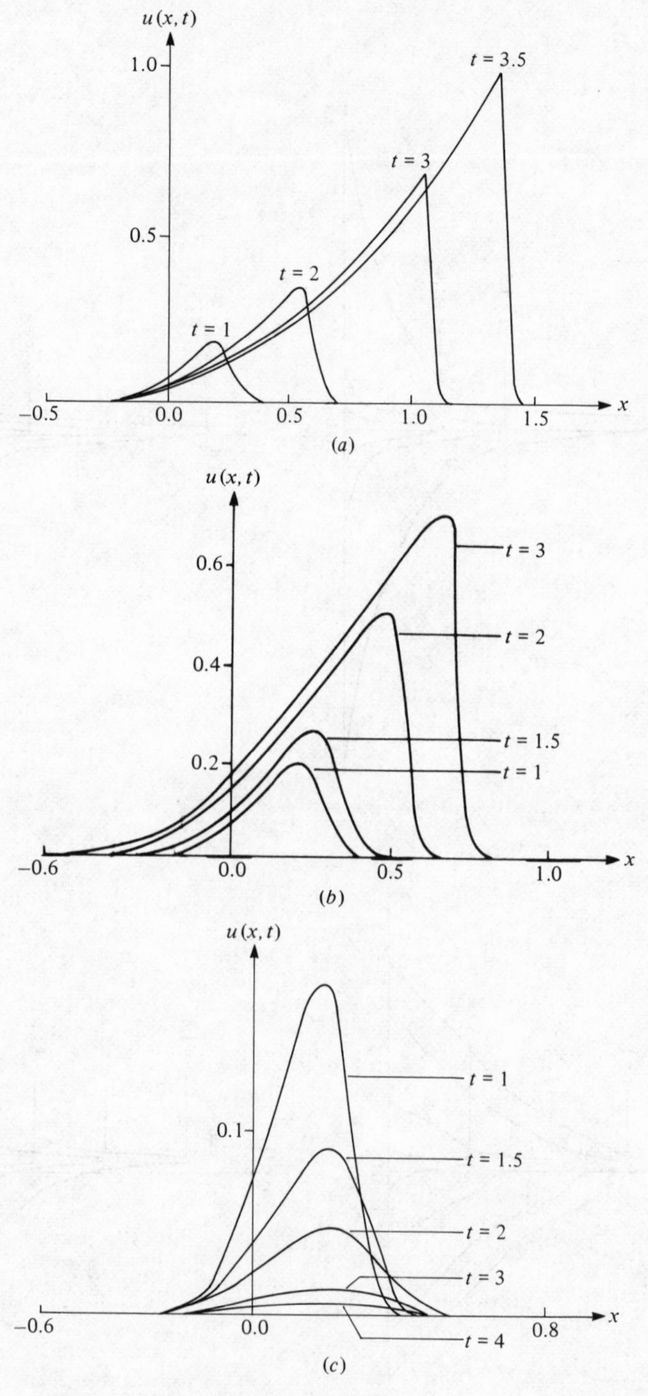

Fig. 5.25. Solution of eq. (5.67) for $\alpha = 1$. (a) $\beta = 0.5, \lambda = -1$. (b) $\beta = 1$, $\lambda = -1$. (c) $\beta = 1$, $\lambda = 1$.

solution grows to a peak somewhere in the middle in a short time, and shows some small persisting wiggles when $\beta > 1$. When $1 \leqslant \alpha < 2$ (figs. 5.25(a)–(b) and 5.26), the solution grows and breaks at the front in a short time. For the case $\alpha = 2$ (fig. 5.27), the solution first decays (implying the dominance of nonlinear convection and diffusion in the early stages) and then grows to break at the front. For $\alpha > 2$ (figs. 5.28–5.30), the negative damping is too small and the solution continuously decays with time.

It is of some interest to compare the special case (5.98) (Lardner and Arya,

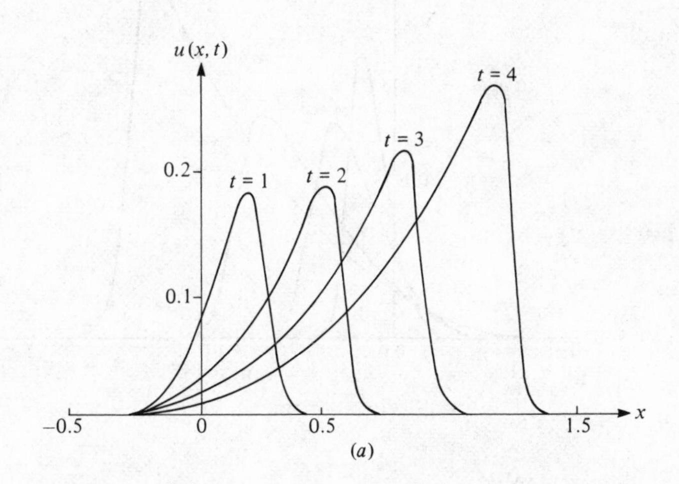

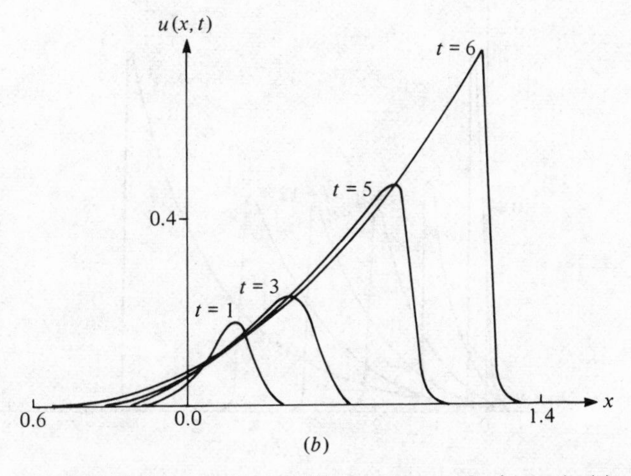

Fig. 5.26. Solution of eq. (5.67) for $1 < \alpha < 2$, $\lambda = -1$. (a) $\alpha = 1.5$, $\beta = 0.5$. (b) $\alpha = 1.5$, $\beta = 1$.

1980) with the corresponding modified K–dV eq. (5.69). Both have the same form of convective and damping terms. Leibovich and Randall have numerically studied an initial value problem for eq. (5.69). They treated a whole class of initial conditions, which give rise to a variety of solitons, differing in number and amplifying or decaying depending on whether $\lambda < 0$

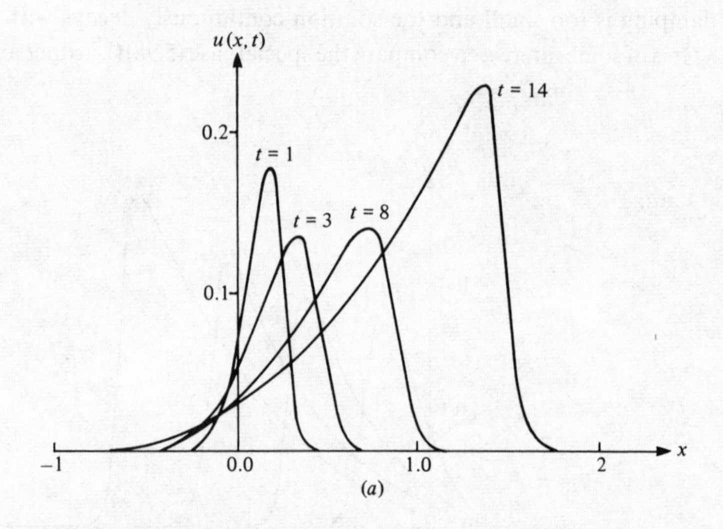

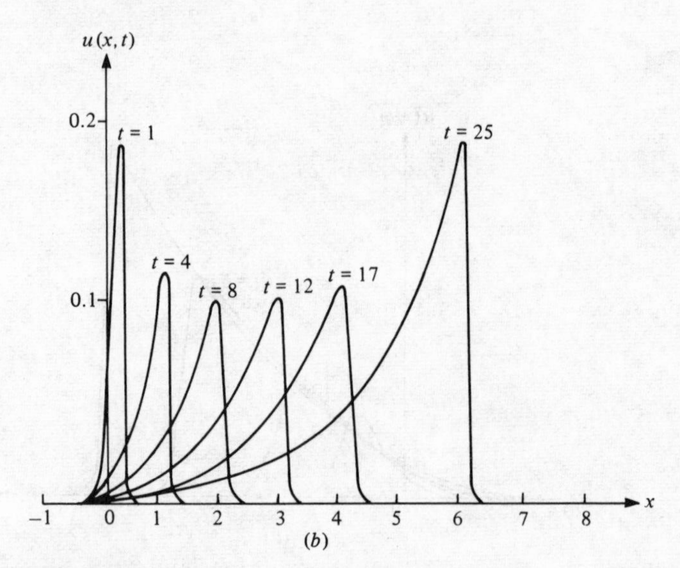

Fig. 5.27. Solution of eq. (5.67) for $\alpha = 2$, $\lambda = -1$. (a) $\beta = 1$. (b) $\beta = 0.5$.

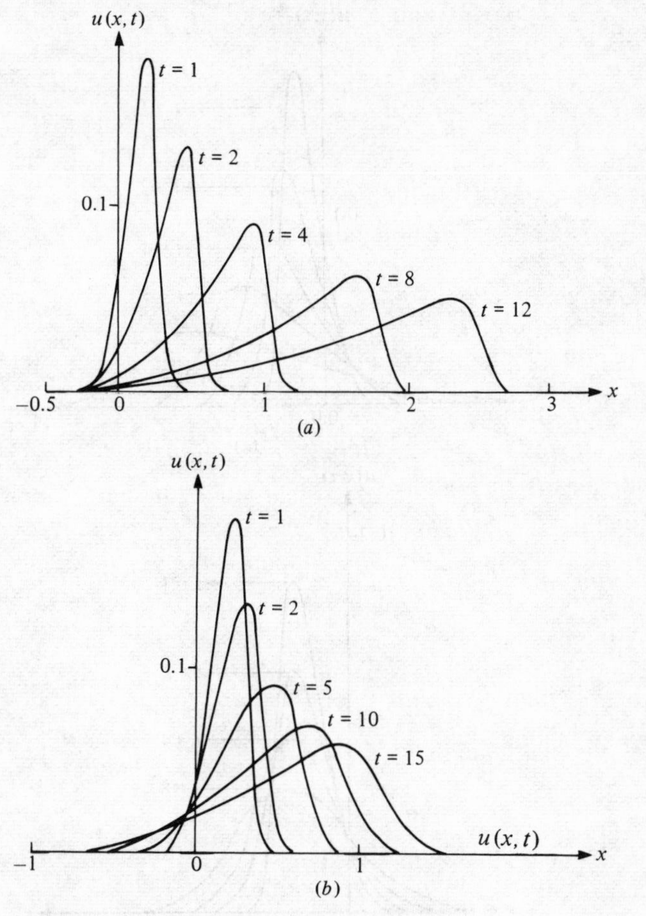

Fig. 5.28. Solution of eq. (5.67) for $2 < \alpha < 3$, $\lambda = -1$. (a) $\alpha = 2.5$, $\beta = 0.5$. (b) $\alpha = 2.5$, $\beta = 1$.

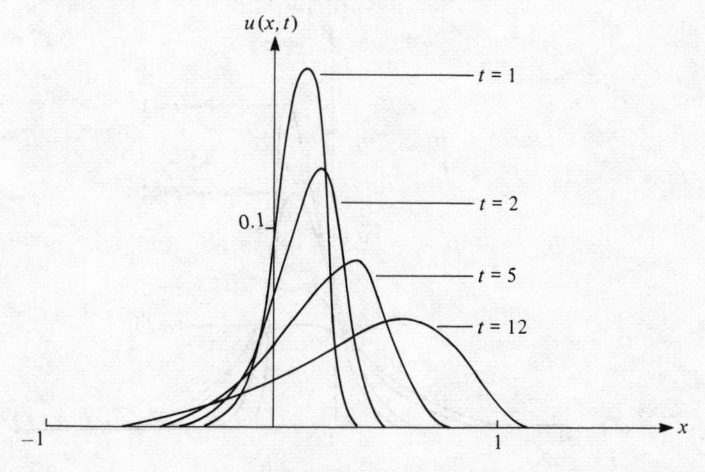

Fig. 5.29. Solution of eq. (5.67) for $\alpha = 3, \beta = 1, \lambda = -1$.

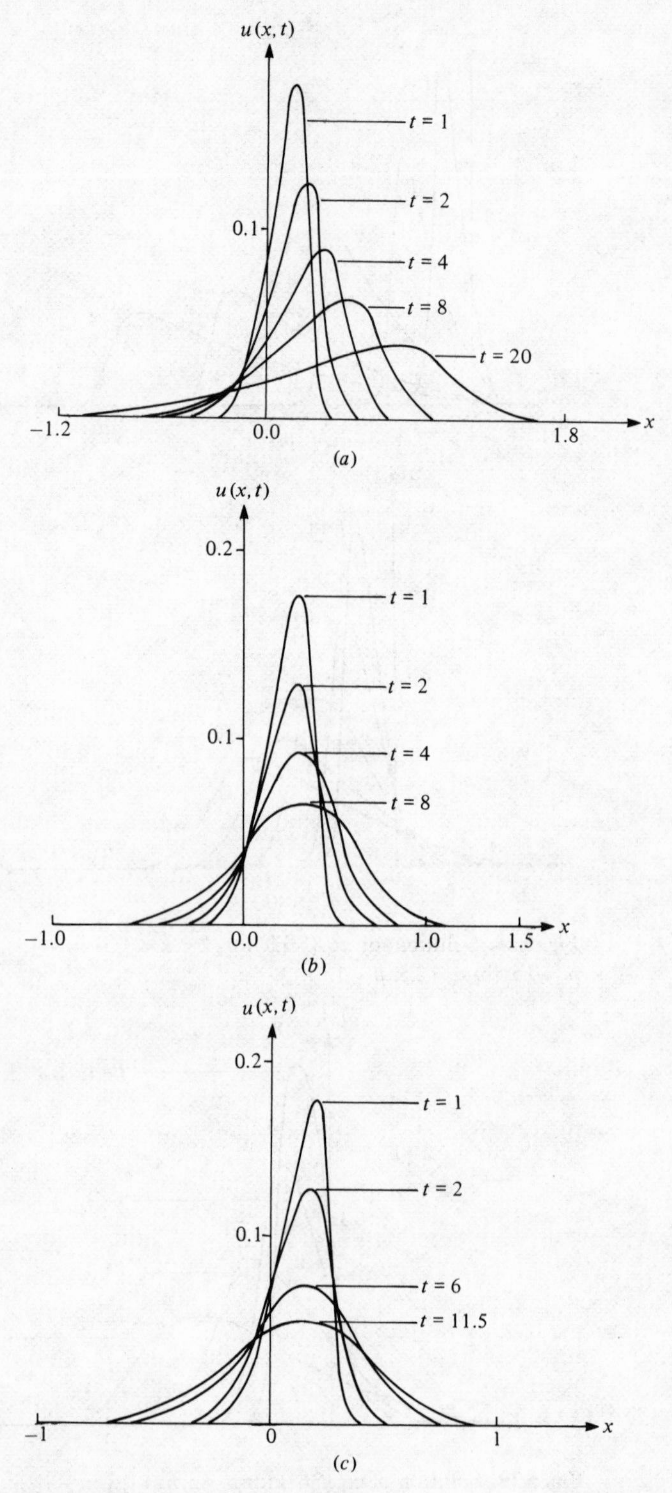

Fig. 5.30. Solution of eq. (5.67) for $\alpha > 3$, $\lambda = -1$. (a) $\alpha = 4$, $\beta = 1$. (b) $\alpha = 4, \beta = 1.5$. (c) $\alpha = 5, \beta = 2$.

or $\lambda > 0$. They discovered three integrals for the entities

$$\left.\begin{aligned} A_r &= \int_{-\infty}^{\infty} u(x,t)\,\mathrm{d}x, \\ E &= \frac{1}{2}\int_{-\infty}^{\infty} u^2(x,t)\,\mathrm{d}x, \\ \bar{x} &= \frac{1}{A_r}\int_{-\infty}^{\infty} xu\,\mathrm{d}x, \end{aligned}\right\} \tag{5.104}$$

which represent area under the wave or its momentum, its energy and centre of gravity, respectively. It is easily checked by direct integration of eq. (5.69) and integration after multiplication by u and x, respectively, that A_r, E and $\bar{x}$ satisfy the following relations:

$$\begin{aligned} A_r &= A_0\,\mathrm{e}^{-\lambda t}, \\ E &= E_0\,\mathrm{e}^{-2\lambda t}, \\ \bar{x} - \bar{x}_0 &= \frac{\mu E_0}{\lambda A_0}(\mathrm{e}^{-\lambda t_0} - \mathrm{e}^{-\lambda t}). \end{aligned} \tag{5.105}$$

The subscript 0 refers to the value of the relevant quantity at $t = t_0$. It is straightforward to check that the integrals (5.105) with $\mu = -1$ exist for the modified Burgers equation (5.98) as well. The main features in the solitary wave study of Leibovich and Randall is the appearance of a trailing shelf and amplification or decay of the wave depending on the sign of λ. They also found a terminal similarity solution for eq. (5.69) for each soliton in isolation. While this solution confirmed the major features (dominant soliton plus shelf), it was not a uniformly valid solution since it failed to satisfy the boundary conditions at $x = \infty$. Eq. (5.98) does not possess a self-similar solution. It shows amplification or decay of the initial profile depending on whether $\lambda < 0$ or $\lambda > 0$ (see figs 5.25(b)–(c)). These numerical solutions satisfy the relations (5.105).

5.8 Non-planar Burgers equation with general nonlinearity

Now we treat the equation

$$u_t + u^\alpha u_x + \frac{ju}{2t} = \frac{\delta}{2}u_{xx}, \quad j = 1, 2 \tag{5.84}$$

in some detail; we had referred to it briefly in sec. 5.7. Since the analysis and numerical results for eq. (5.84) are rather similar to those for eq. (5.67), we shall not dilate upon them. Eq. (5.84), however, admits considerable analysis; indeed we find here a single parameter family of exact solutions for $j = 1, 2$ for specific values of α. The chief purpose of the present section is to fortify our claim that the Euler–Painlevé transcendents solving eq. (5.81) characterise GBEs in the manner the Painlevé equations represent K–dV type of equations.

It is easy to check that eq. (5.84) admits the self-similar form

$$u = t^{-1/2\alpha} f(\eta), \quad \eta = x(2\delta t)^{-1/2}, \tag{5.106}$$

resulting in the nonlinear ODE

$$f'' - 2^{3/2}\delta^{-1/2} f^{\alpha} f' + 2\eta f' + \frac{2(1-\alpha j)}{\alpha} f = 0 \tag{5.107}$$

where the prime denotes differentiation with respect to η. The transformation

$$H = \delta^{1/2} f^{-\alpha} \tag{5.108}$$

changes eq. (5.107) into

$$HH'' - \frac{\alpha+1}{\alpha} H'^2 + 2\eta H H' - 2(1-\alpha j)H^2 - 2^{3/2} H' = 0. \tag{5.109}$$

Eq. (5.109) is a special case of eq. (5.81) with $a = -(\alpha+1)/\alpha$, $f(\xi) = 2\xi$, $g(\xi) = -2(1-\alpha j)$, $b = -2^{3/2}$ and $c = 0$. The solution of eq. (5.109) can be written in the form of a Taylor series:

$$H(\eta) = \sum_{n=0}^{\infty} a_n \eta^n \tag{5.110}$$

where the coefficients a_n can be found by substitution into eq. (5.109) as

$$a_2 = \frac{1}{a_0}\left[\frac{\alpha+1}{2\alpha} a_1^2 + (1-\alpha j) a_0^2 + 2^{1/2} a_1\right], \tag{5.111}$$

$$\begin{aligned} a_{k+2} = \frac{2}{(k+1)(k+2)a_0} \Bigg\{ & \frac{\alpha+1}{2\alpha}(k+1) a_1 a_{k+1} \\ & + (1-\alpha j - k) a_0 a_k + 2^{1/2}(k+1) a_{k+1} \\ & + \sum_{i=1}^{k} \Bigg[-\tfrac{1}{2}(k+2-i)(k+1-i) a_i a_{k+2-i} \\ & + \frac{\alpha+1}{2\alpha}(i+1)(k+1-i) a_{i+1} a_{k+1-i} + (1-\alpha j) a_i a_{k-i} \\ & - (k-i) a_i a_{k-i} \Bigg] \Bigg\}, \quad k = 1, 2, \ldots. \end{aligned} \tag{5.112}$$

Thus, in general, we have a two-parameter a_0, a_1 family of solutions. However, for the special choice $\alpha = 1/(j+1), j = 0, 1, 2$, and $a_1 = -2^{3/2}\alpha/(\alpha+1)$, the coefficient a_2 in eq. (5.111) depends only on a_0 and we have a single parameter family of solutions. This series can in fact be summed and the solution written in terms of exponential and erfc functions, as we shall show directly (see eq. (5.120) below). The free parameter a_0 in this special case could be either the amplitude or the Reynolds number $R = \frac{1}{\delta}\int_{-\infty}^{\infty} u\,\mathrm{d}x$ of the profile, as for the standard Burgers equation.

We find the asymptotic solution of eq. (5.107) for large $|\eta|$ under the conditions that $f \to 0$ as $\eta \to \pm\infty$.

The linearised form of eq. (5.107), namely

$$f'' + 2\eta f' + \frac{2(1-\alpha j)}{\alpha} f = 0, \tag{5.113}$$

has the solution

$$f(\eta) = A\,\mathrm{e}^{-\eta^2} H_\nu(\eta), \quad \eta > 0, \tag{5.114a}$$

$$f(\eta) \sim \frac{B\pi^{1/2}}{\Gamma(-\nu)} |\eta|^{j-1/\alpha}, \quad \text{for large negative } \eta, \tag{5.114b}$$

provided $\alpha j < 1$. Here, $\nu = 1/\alpha - (j+1)$ and $H_\nu(\eta)$ is the Hermite function of order ν; A and B are the amplitude parameters. Thus, the linear solution decays exponentially as $\eta \to +\infty$ and algebraically as $\eta \to -\infty$.

We now pose the boundary value or connection problem for eq. (5.107):

$$f'' - 2^{3/2}\delta^{-1/2} f^\alpha f' + 2\eta f' + \frac{2(1-\alpha j)}{\alpha} f = 0, \tag{5.107}$$

$$f \sim A\mathrm{e}^{-\eta^2} H_\nu(\eta), \quad \eta \uparrow \infty, \tag{5.115a}$$

$$f \to 0, \quad \eta \downarrow -\infty \tag{5.115b}$$

and

$$|f| < \infty, \quad -\infty < \eta < \infty. \tag{5.116}$$

The study of this connection problem is rendered easier by the following analysis which gives some special exact solutions and identifies the ranges of parameter α for which the solutions vanish either at $\eta = \to -\infty$ or at a finite value of $\eta = \eta_0$, say.

For $\alpha = 1/(j+1)$, eq. (5.107) can be written as

$$f + \eta f' + \tfrac{1}{2} f'' = \left(\frac{2}{\delta}\right)^{1/2} f^\alpha f' \tag{5.117}$$

which can be immediately integrated. Using the conditions that f and f'

tend to zero as $\eta \to +\infty$, we have

$$\eta f + \tfrac{1}{2} f' = \frac{1}{\alpha+1}\left(\frac{2}{\delta}\right)^{1/2} f^{\alpha+1}. \tag{5.118}$$

Writing $f^{-\alpha} = G$ in eq. (5.118), we have

$$G' - 2\alpha\eta G = -\frac{2\alpha}{\alpha+1}\left(\frac{2}{\delta}\right)^{1/2}. \tag{5.119}$$

This equation can again be integrated so that we get

$$G = \left[C - \frac{2}{\alpha+1}\left(\frac{2\alpha}{\delta}\right)^{1/2} \int_0^{\alpha^{1/2}\eta} \mathrm{e}^{-t^2}\,\mathrm{d}t \right] \mathrm{e}^{\alpha\eta^2}. \tag{5.120}$$

The solution (5.120), for the original variable f, can be written as

$$f(\eta) = \mathrm{e}^{-\eta^2}\left[C - \frac{2}{\alpha+1}\left(\frac{2\alpha}{\delta}\right)^{1/2} \int_0^{\alpha^{1/2}\eta} \mathrm{e}^{-t^2}\,\mathrm{d}t \right]^{-1/\alpha}, \tag{5.121}$$

where $C = f^{-\alpha}(0)$. The solution $u = t^{-1/2\alpha} f(\eta)$ of eq. (5.84) with f as in eq. (5.121), we believe, is new. It holds for $\alpha = 1/(j+1)$, that is, for $j = 1$, $\alpha = \frac{1}{2}$, and $j = 2, \alpha = \frac{1}{3}$, and generalises the exact single hump solution of the standard Burgers equation with $j = 0$, $\alpha = 1$. We further note that, for $\alpha = 1/j$, $j \neq 0$, f equal to an arbitrary constant ($\neq 0$) is another solution of eq. (5.107).

If we write $f^{\alpha} = F$ in eq. (5.107), we have

$$\tfrac{1}{2} F F'' - \frac{\alpha-1}{2\alpha} F'^2 + (1 - \alpha j) F^2 + \eta F F' - \left(\frac{2}{\delta}\right)^{1/2} F^2 F' = 0. \tag{5.122}$$

Integrating eq. (5.122) with respect to η from $-\infty$ to $+\infty$ and assuming that both F and F' tend to zero as η tends to $-\infty$ or $+\infty$, we get

$$(2\alpha j - 1)\int_{-\infty}^{\infty} F^2\,\mathrm{d}\eta = \left(\frac{1-2\alpha}{\alpha}\right)\int_{-\infty}^{\infty} F'^2\,\mathrm{d}\eta. \tag{5.123}$$

This equality yields the following results:

(i) $j = 0$. The ratio

$$r = \frac{\displaystyle\int_{-\infty}^{\infty} F^2\,\mathrm{d}\eta}{\displaystyle\int_{-\infty}^{\infty} F'^2\,\mathrm{d}\eta} = -(1-2\alpha)/\alpha > 0 \quad \text{if } \alpha > \tfrac{1}{2}. \tag{5.124}$$

Table 5.25. *Single hump, monotonic and diverging solutions of* (5.107) *and* (5.115a)

Behaviour at left boundary	$j=0$	$j=1$	$j=2$
Solutions vanishing at $\eta=-\infty$	$\alpha \geqslant 1$	$\alpha=\frac{1}{2}$	$\frac{1}{3}\leqslant\alpha<\frac{1}{2}$
Solutions vanishing at $\eta=\eta_0$	$\frac{1}{2}<\alpha<1$	—	$\frac{1}{4}<\alpha<\frac{1}{3}$
Solutions monotonically approaching a constant at $\eta=-\infty$	—	$\alpha=1$	$\alpha=\frac{1}{2}$
Solutions diverging to infinity at $\eta=-\infty$	—	$\alpha>1$	$\alpha>\frac{1}{2}$

Therefore, the single hump solutions vanishing at $\eta=\pm\infty$ exist only if $\alpha>\frac{1}{2}$ (see, however, the discussion following the case (iii)).

(ii) $j=1$. The only valid choice in this case is $\alpha=\frac{1}{2}$ and this value corresponds to the exact solution (5.121) noted earlier.

(iii) $j=2$. In this case, the ratio $r=(1-2\alpha)/\alpha(4\alpha-1)>0$ if $\frac{1}{4}<\alpha<\frac{1}{2}$. This is the range in which the single hump solutions vanishing at $\eta=\pm\infty$ exist.

If the solutions starting from $\eta=+\infty$ according to eq. (5.115a) vanish not at $\eta=-\infty$ but at a finite $\eta=\eta_0$, then the above results do not hold. At $\eta=\eta_0$, $f'>0$. Integrating eq. (5.107) from $\eta=\eta_0$ to $\eta=\infty$, we find that

$$\frac{\alpha(j+1)-1}{\alpha}\int_{\eta_0}^{\infty} f\,\mathrm{d}\eta=-\tfrac{1}{2}f'(\eta_0)<0. \qquad (5.125)$$

Since $f>0$ for $\eta_0<\eta<\infty$, eq. (5.125) implies that $\alpha<1/(j+1)$. Thus, single hump solutions vanishing at $\eta=+\infty$ and $\eta=\eta_0$, a finite point on the left, exist only if $\alpha<1/(j+1)$. Combining this result with those in (i)–(iii), we find that single hump solutions vanishing at $\eta=+\infty$ and at either $\eta=-\infty$ or $\eta=\eta_0$ exist if $1/(j+2)<\alpha<1/(j+1)$, $j=0,2$. We note in passing that the condition that a maximum exists for the single hump solution, requiring $f'=0$, $f''<0$ at $\eta=\eta_{\max}$, say, leads to the inequality $\alpha j<1$ (see eq. (5.107)). However, this condition is too lax in comparison with those which naturally arise from the consideration of eqs. (5.123) and (5.125).

Table 5.25 summarises the nature of the solution for different values of j and α, following from the above discussion and the numerical results.

The numerical solution of eq. (5.107) was carried out subject to (5.115a), starting the integration from a large positive value of $\eta\sim4$ where the values of f and f' are $O(10^{-5})$. We proceeded towards decreasing values of η. This was done first for the ranges of parameters for which the single hump solutions vanishing at $\eta = \pm\infty$ exist. From the numerical solution the values of $f(0)$ and $f'(0)$ were obtained, $H(0)$ and $H'(0)$ were then found using eq. (5.108), and hence the Taylor series (5.110) was computed. The series solution so obtained agreed very closely with the numerical solution of the connection problem for eq. (5.107). It was found to be accurate to seven decimal places in single precision arithmetic for all η lying between $-\infty$ and $+\infty$. The series solution was analytically continued as its convergence slowed down. We further confirmed the accuracy of the series and numerical solutions by comparing them with the exact solutions (5.121) for the special values $\alpha = 1/(j+1)$. In the similarity range of the parameter α, for various geometries, it was found that the solution of the connection problem exists for all values of the amplitude parameter A. Thus, it is the nonlinearity exponent α which determines the existence (or otherwise) of the single hump solutions. This is in contrast to the case of the damped GBE (5.67) for which the self-similar solutions, for given α in the permissible range, are restricted by the magnitude of the amplitude parameter.

The nature of the solution as given in Table 5.25 was fully verified numerically.

The numerical solution of the PDE (5.84) was also carried out by pseudo-spectral and implicit schemes to visualise the evolution of the self-similar solutions from a class of 'reasonable' continuous and discontinuous initial conditions vanishing at $\eta = \pm\infty$. This part of the programme is entirely analogous to that for eq. (5.67). Since the latter has been discussed in great detail in sec. 5.7, we skip the details relevant to eq. (5.84) and refer the reader to the paper by Sachdev and Nair (1987).

We conclude this section with the hope that our conjecture regarding the role of the Euler–Painlevé equation (5.81) will be confirmed for other GBEs besides eqs. (5.67) and (5.84) which we have dealt with here.

References

Abdelkader, M. A. (1982). Travelling wave solutions for a generalised Fisher equation. *J. Math. Anal. Appl.* **85**, 287–90.

Ablowitz, M. J., Ramani, A. and Segur, H. (1980). A connection between nonlinear evolution equations and ordinary differential equations of *P*-type. I. *J. Math. Phys.* **21**, 715–21.

Ablowitz, M. J. and Zeppetella, A. (1979). Explicit solutions of Fisher's equation for a special wave speed. *Bull. Math. Biol.* **41**, 835–40.

Abramowitz, M. and Stegun, I. A. (eds.) (1964). *Handbook of mathematical functions.* National Bureau of Standards, Washington D. C.

Ames, W. F. (1965). *Nonlinear partial differential equations in engineering*, Vol. I. Academic Press, New York.

Ames, W. F. (1972). *Nonlinear partial differential equations in engineering*, Vol. II. Academic Press, New York.

Anderson, R. L. and Ibragimov, N. H. (1979). *Lie–Bäcklund transformations in applications.* SIAM, Philadelphia.

Arakawa, A. (1966). Computational design for long-term numerical integration of the equations of fluid motion. Two-dimensional incompressible flow. Part I. *J. Comp. Phys.* **1** (1966), 119–43.

Atkinson, F. V. and Peletier, L. A. (1974). Similarity solutions of the nonlinear diffusion equation. *Arch. Rat. Mech. Anal.* **54**, 373–92.

Barenblatt, G. I. (1979). *Similarity, self-similarity and intermediate asymptotics.* Consultants Bureau, New York.

Barenblatt, G. I. and Zel'dovich, Y. B. (1972). Self-similar solutions as intermediate asymptotics. *Ann. Rev. Fluid Mech.* **4**, 285–312.

Bateman, H. (1915). Some recent researches on the motion of fluids, *Mon. Weather Rev.* **43**, 163–70.

Bender, C. M. and Orszag, S. A. (1978). *Advanced mathematical methods for scientists and engineers.* McGraw-Hill, New York.

Benton, E. R. and Platzman, G. W. (1972). A table of solutions of one-dimensional Burgers equation. *Quart. Appl. Math.* **30**, 195–212.

Berryman, J. G. (1977). Evolution of a stable profile for a class of nonlinear diffusion equations with fixed boundaries. *J. Math. Phys.* **18**, 2108–15.

Berryman, J. G. and Holland, C. J. (1978). Evolution of a stable profile for a class of nonlinear diffusion equations II. *J. M*ath. *Phys.* **19**, 2476–80.

Berryman, J. G. and Holland, C. J. (1982). Asymptotic behaviour of the nonlinear diffusion equation $n_t = (n^{-1} n_x)_x$. *J. Math. Phys.* **23**, 983–7.

Bertsch, M. (1982). Asymptotic behaviour of solutions of a nonlinear diffusion equation. *SIAM J. Appl. Math.* **42**, 66–76.

Bhatnagar, P. L. (1979). *Nonlinear waves in one-dimensional dispersive systems.* Oxford University Press, Delhi.

Birkhoff, G. and Rota, G. C. (1978). *Ordinary differential equations* (3rd edition). John Wiley, New York.

Blackstock, D. T. (1964). Thermoviscous attenuation of plane, periodic finite-amplitude sound waves. *J. Acoust. Soc. Amer.* **36**, 534–42.

Blasius, H. (1908). Grenzschichten in Flüssigkeiten mit kleiner Reibung. *Z. Math. u. Phys.* **56**, 1–37.

Bluman, G. W. and Cole, J. D. (1969). The general similarity solution of the heat equation. *J. Math. Mech.* **18**, 1025–42.

Broer, L. J. F. and Schuurmans, M. F. H. (1970). On a simple wave approximation. *J. Eng. Maths.* **4**, 305–18.

Burgers, J. M. (1940). Application of a model system to illustrate some points of the statistical theory of turbulence. *Proc. Roy Neth. Acad. Sci. Amst.* **43**, 2–12.

Burgers, J. M. (1950). The formation of vortex sheets in a simplified type of turbulent motion. *Proc. Roy. Neth. Acad. Sci. Amst.* **53**, 122–33, see also pp. 247–60, 393–406, 718–42.

Burgers, J. M. (1974). *The nonlinear diffusion equation.* Reidel, Dordrecht.

Canosa, J. (1969). Diffusion in nonlinear multiplicative media. *J. Math. Phys.* **10**, 1862–8.

Canosa, J. (1973). On a nonlinear diffusion equation describing population growth. *IBM J. Res. Develop.* **17**, 307–13.

Case, K. M. and Chiu, C. S. (1969). Burgers' turbulence models. *Phys. Fluids* **12**, 1799–1808.

Chester, W. (1977). Continuous transformations and differential equations. *J. Inst. Math. Appl.* **19**, 343–76.

Chu, C. W. (1965). A class of reducible systems of quasi-linear partial differential equations. *Quart. Appl. Math.* **23**, 275–8.

Coddington, E. A. and Levinson, N. (1955). *Theory of ordinary differential equations.* McGraw-Hill, New York.

Cole, J. D. (1951). On a quasi-linear parabolic equation occurring in aerodynamics. *Quart. Appl. Math.* **9**, 225–36.

Colton, D. (1974). Integral operators and reflection principles for parabolic equations in one space variable. *J. Diff. Eqns.* **15**, 551–9.

Colton, D. (1975). Complete families of solutions for parabolic equations with analytic coefficients. *SIAM. J. Math. Anal.* **6**, 937–47.

Colton, D. (1976). The approximation of solutions to initial boundary value problems for parabolic equations in one space variable. *Quart. Appl. Math.* **34**, 377–86.

Cooley, J. W., Lewis, P. A. W. and Welsh, P. D. (1969). The finite Fourier transform. *IEEE Trans. Education* **E-12**, 27–34.

Copson, E. T. (1975). *Partial differential equation.* Cambridge University Press.

Courant, R. and Friedrichs, K. O. (1948). *Supersonic flow and shock waves.* Interscience, New York.

Courant, R. and Hilbert, D. (1962). *Methods of mathematical physics*, Vol. II. Interscience, New York.

Crank, J. (1975). *The mathematics of diffusion.* Clarendon Press, Oxford.

Crighton, D. G. (1979). Model equations of nonlinear acoustics. *Ann. Rev. Fluid Mech.* **11**, 11–23.

Crighton, D. G. and Scott, J. F. (1979). Asymptotic solution of model equations in nonlinear acoustics. *Phil. Trans. Roy. Soc.* **A292**, 101–34.

Dafermos, C. M. (1973). Solution of the Riemann Problem for a class of hyperbolic systems of conservation laws by the viscosity method. *Arch. Rat. Mech. Anal.* **52**, 1–9.

Dafermos, C. M. (1974). Quasilinear hyperbolic systems that result from conservation laws: in Leibovich and Seebass (1974), 82–102.

Davis, H. T. (1962). *Introduction to nonlinear differential and integral equations.* Dover Publications, New York.

Diekmann, O., Hilhorst, D. and Peletier, L.A. (1980). A singular boundary value problem arising in a pre-breakdown gas discharge. *SIAM J. Appl. Maths.* **39**, 48–66.

Dorodnitsyn, V. A. (1982). On invariant solutions of the equation of nonlinear heat conduction with a source. *USSR Comp. Math. Phys.* **22**, 115–22.

Douglas, J. and Jones, B. F. (1963). On predictor–corrector methods for nonlinear parabolic differential equations. *J. Soc. Ind. and Appl. Math.* **11**, 195–204.

Drake, J. R. and Berryman, J. G. (1977). Theory of nonlinear diffusion of plasma across the magnetic field of a toroidal multipole. *Phys. Fluids.* **20**, 851–7.

Drake, J. R., Greenwood, J. R., Navratil, G. A. and Post, R. S. (1977). Diffusion coefficient scaling in the Wisconsin levitated octupole. *Phys. Fluids.* **20**, 148–55.

Earnshaw, S. (1858). On the mathematical theory of sound. *Phil. Trans. Roy. Soc.* **A150**, 133–48.

Fay, R. D. (1931). Plane sound waves of finite amplitude. *J. Acoust. Soc. Amer.* **3**, 222–41.

Fisher, R. A. (1936). The wave of advance of advantageous genes. *Ann. of Eugen.* **7**, 355–69.

Fokas, A. S. and Yortsos, Y. C. (1982). On the exactly solvable equation $S_t = [(\beta S + \gamma)^{-2} S_x]_x + \alpha(\beta S + \gamma)^{-2} S_x$ occurring in two-phase flow in porous media. *SIAM J. Appl. Math.* **42**, 318–32.

Fornberg, B. and Whitham, G. B. (1978). A numerical and theoretical study of certain nonlinear wave phenomena. *Phil. Trans. Roy. Soc.* **A289**, 373–404.

Forsyth, A. R. (1906). *Theory of differential equations*, Vol. 6. Cambridge University Press.

Friedman, A. (1964). *Partial differential equations of parabolic type.* Prentice-Hall, Englewood Cliffs, New Jersey.

Fubini-Ghiron, E. (1935). Anomalie nella propagazione di onde acustiche di grande ampiezza. *Alta Freq.* **4**, 530–81.

Gazdag, J. (1973). Numerical convective schemes based on accurate computation of space derivatives. *J. Comp. Phys.* **13**, 100–13.

Gazdag, J. and Canosa, J. (1974). Numerical solution of Fisher's equation. *J. Appl. Prob.* **11**, 445–57.

Gelfand, I. M. (1959). Some problems of the theory of quasi-linear equations. *Uspekhi Mat. Nauk*, **14**, 82–158.

Grinberg, G. A. (1969). On the temperature or concentration fields produced inside an infinite or finite domain by moving surfaces at which the temperature or concentration are given as functions of time. *J. Appl. Math. and Mech.* (PMM) **33**, 1021–9.

Grundy, R.E. (1979). Similarity solutions of the nonlinear diffusion equation. *Quart. Appl. Math.* **37**, 259–80.

Guderley, G. (1942). Starke kugelige und zylindrische Verdichtungsstösse in der Nähe des Kugelmittelpunktes bzw. der Zylinderachse, *Luftfahrtforschung* **19**, 302–12.

Gurbatov, S. N., Saichev, A. I. and Yakushkin, I. G. (1983). Nonlinear waves and one-dimensional turbulence in non-dispersive media. *Sov. Phys. Usp.* **26**, 857–76.

Hagstrom, T. and Keller, H. B. (1986). The numerical calculations of travelling wave solutions of nonlinear parabolic equations (Preprint).

Hastings, S. P. and McLeod, J. B. (1980). A boundary value problem associated with the second Painlevé transcendent and the Korteweg–de Vries equation. *Arch. Rat. Mech. Anal.* **73**, 31–51.

Hilhorst, D. (1982). A nonlinear evolution problem arising in the physics of ionized gases. *SIAM J. Math. Anal.* **13**, 16–39.

Hille, E. (1969). *Lectures on ordinary differential equations.* Addison-Wesley, Reading, Mass.

Hille, E. (1970). Some aspects of Thomas–Fermi equation. *J. Analyse Math.* **23**, 147–70.

Hirota, R. (1971). Exact solution of the Korteweg–de Vries equation for multiple collisions of solitons. *Phys. Rev. Lett.* **27**, 1192–4.

Holland, C. J. (1977). On the limiting behaviour of Burgers' equation. *J. Math. Anal. Appls.* **57**, 156–60.

Hopf, E. (1950). The partial differential equation $u_t + uu_x = \mu u_{xx}$. *Comm. Pure Appl. Math.* **3**, 201–30.

Humi, M. (1977). Invariant solutions for a class of diffusion equations. *J. Math. Phys.* **18**, 1705–8.

Ince, E. L. (1956). *Ordinary differential equations.* Dover Publications, New York.

Itaya, N. (1976). A survey on the generalised Burgers equation with a pressure model term. *J. Math. Kyoto Univ.* **16**, 223–40.

Jain, M. K. (1979). *Numerical solution of differential equations.* Wiley Eastern, New Delhi.

Jeng, D. T. and Meecham, W. C. (1972). Solutions of forced Burgers equation. *Phys. Fluids* **15**, 504–6.

Johnson, R. S. (1970). A nonlinear equation incorporating damping and dispersion. *J. Fluid Mech.* **42**, 49–60.

Johnson, R. S. (1972). Shallow water waves on viscous fluid – the undular bore. *Phys. Fluids* **15**, 1693–9.

Jones, C. W. (1953). On reducible nonlinear differential equations occurring in mechanics. *Proc. Roy. Soc.* **A217**, 327–43.

Kamenomostskaya, S. (1973). Asymptotic behaviour of the solution of the filtration equation. *Israel J. Math.* **14**, 76–87.

Kamenomostskaya, S. (1978). Source type solutions for equations of non-stationary filtration. *J. Math. Anal. Appl.* **64**, 263–76.

Kametaka, Y. (1976). On the nonlinear diffusion equation of Kolmogorov–Petrovskii–Piskunov type. *Osaka J. Math.* **13**, 11–66.

Kamin, S. and Rosenau, P. (1981). Propagation of thermal waves in an inhomogeneous medium. *Comm. Pure and Appl. Math.* **34**, 831–52.

Kamke, E. (1943). *Differential Gleichungen: Lösungsmethoden und Lösungen.* Akademische Verlagsgesellschaft, Leipzig.

Karabutov, A. A. and Rudenko, O. V. (1976). Excitation of nonlinear acoustic waves by surface absorption of laser radiation. *Sov. Phys. Tech. Phys.* **20**, 920–2.

Karpman, V. I. (1975). *Nonlinear waves in dispersive media.* Pergamon, Oxford.

Kevorkian, J. and Cole, J. D. (1981). *Perturbation methods in applied mathematics.* Springer-Verlag, New York.

Khusnytdinova, N. V. (1967). The limiting moisture profile during infiltration into a homogeneous soil. *J. Appl. Math. and Mech.* (*PMM*) **31**, 783–9.

Kingston, J. G. and Rogers, C. (1982). Reciprocal Bäcklund transformations of conservation laws. *Phys. Lett.* **92A**, 261–4.

Kingston, J. G. and Rogers, C. and Woodall, D. (1984). Reciprocal auto-Bäcklund transformations. *J. Phys. A: Mathematical and General* **17**, L35–8.

Kochina, N. N. (1961). On periodic solutions of Burgers equation. *J. Appl. Math. and Mech.* (*PMM*) **25**, 1597–607.

Kolmogoroff, A., Petrovsky, I. and Piscounov, N. (1937). Etude de l'équation de la diffusion avec croissance de la quantité de matière et son application à un problèm biologique. *Bull. de l'Univ. d'Etat à Moscou* (*Ser. Inter.*) **A1**, 1–25.

Kraut, E. A. (1964). The uniqueness of weak solutions of the one-dimensional scalar analog to the Navier–Stokes equations. *J. Math. Phys.* **5**. 1290–2.

Lagerstrom, P. A., Cole, J. D. and Trilling, L. (1949). *Problems in theory of viscous compressible fluids.* Monograph, Calif. Inst. Tech. 232 pages.

Lardner, R. W. (1976). The development of shock waves in nonlinear viscoelastic media. *Proc. Roy. Soc.* **A347**, 329–44.

Lardner, R. W. (1986). Third order solutions of Burgers equation. *Quart. Appl. Math.* **44**, 293–302.

Lardner, R. W. and Arya, J. C. (1980). Two generalisations of Burgers equation. *Acta Mech.* **37**, 179–90.

Larsen, D. (1978). Transient bounds and time-asymptotic behaviour of solutions. *SIAM J. Appl. Math.* **34**, 93–103.

Lax, P. D. (1957). Hyperbolic systems of conservation laws II. *Comm. Pure Appl. Math.* **10**, 537–66.

Lees, M. (1966). A linear three-level difference scheme for quasilinear parabolic equations. *Math. Comp.* **20**, 516–22.

Lefschetz, S. (1977). *Differential equations: Geometric theory*. Dover Publications, Inc., New York.

Lehnigk, S. H. (1976a). Conservative similarity solutions of the one-dimensional autonomous parabolic equation. *J. Appl. Math. Phys.* (*ZAMP*) **27**, 385–91.

Lehnigk, S. H. (1976b). A class of conservative diffusion processes with delta function initial conditions. *J. Math. Phys.* **17**, 973–6.

Leibovich, S. and Randall, J. D. (1979). On soliton amplification. *Phys. Fluids* **22**, 2289–95.

Leibovich, S. and Seebass, A. R. (eds) (1974). *Nonlinear waves*. Cornell University Press, London. (see in particular Chapter 4 by Leibovich and Seebass).

Lesser, M. B. and Crighton, D. G. (1975). Physical acoustics and the method of matched asymptotic expansions. In *Physical acoustics*, ed. W. P. Mason and R. N. Thurston, Vol. 11. Academic Press, New York.

Levine, L. E. (1972). Two-dimensional unsteady self-similar flows in gas-dynamics. *ZAMM* **52**, 441–60.

Lie, S. (1891). Vorlesungen über differentialgleichungen mit bekannten infinitesimalen Transformationen, Leipzig, reprinted by Chelsea Publishing Company, New York, 1967.

Lighthill, M. J. (1956). Viscosity effects in sound waves of finite amplitude. In Surveys in Mechanics, ed. G. K. Batchelor and R. M. Davies, 250–351. Cambridge University Press.

Lighthill, M. J. (1978). *Waves in fluids*. Cambridge University Press.

Luikov, A. V. (1966). *Heat and mass transfer in capillary porous bodies*. Pergamon Press, Oxford.

Luning, C. D. and Perry, W. L. (1981). Convergence of Berryman's iterative method for some Emden-Fowler equations. *J. Math. Phys.* **22**, 1591–5.

Meek, P. C. and Norbury, J. (1982). Two-stage two-level finite difference schemes for nonlinear parabolic equations. *IMA J. Num. Anal.* **2**, 335–56.

Mendousse, J. S. (1953). Nonlinear dissipative distortion of progressive sound waves at moderate amplitudes. *J. Acoust. Soc. Amer.* **25**, 51–4.

Miles, J. W. (1978). On the second Painlevé transcendent. *Proc. Roy. Soc.* **A361**, 277–91.

Mitchell, A. R. and Griffiths, D. F. (1980). *The finite difference method in partial differential equations*. Wiley, New York.

Munier, A., Burgen, J. R., Gutierrez, J., Fijalkow, E. and Feix, M. R. (1981). Group transformations and the nonlinear heat diffusion equation. *SIAM J. Appl. Math.* **40**, 191–207.

Murray, J. D. (1968). Singular perturbations of a class of nonlinear hyperbolic and parabolic equations. *J. Math. Phys.* **47**, 111–33.

Murray, J. D. (1970a). Perturbation effects on the decay of discontinuous solutions of nonlinear first order wave equations. *SIAM J. Appl. Math.* **19**, 135–60.

Murray, J. D. (1970b). On the Gunn-effect and other physical examples of perturbed conservation equations. *J. Fluid Mech.* **44**, 315–46.

Murray, J. D. (1973) On Burgers' model equation for turbulence. *J. Fluid Mech.* **59**, 263–79.

Nayfeh, A. H. (1973). *Perturbation methods.* Wiley, New York.

Newman, W. I. (1983). Nonlinear diffusion: self-similarity and travelling waves. *Pure and Appl. Geoph.* **121**, 417–41.

Nimmo, J. J. C. and Crighton, D. G. (1982). Bäcklund transformations for nonlinear parabolic equations: the general results. *Proc. Roy. Soc.* **A384**, 381–401.

Okamura, M. and Kawahara, T. (1983). Steady solutions of forced Burgers equation. *J. Phys. Soc. Japan* **52**, 3800–6.

Okuda, H. and Dawson, J. M. (1972). Numerical simulation of plasma diffusion in three dimensions. *Phys. Rev. Lett.* **28**, 1625–31.

Oleinik, O. A. and Kruzhkov, S. N. (1961) Quasi-linear second-order parabolic equations with many independent variables. *Russian Math. Surveys* **16**, 105–46.

Ovsiannikov, L. V. (1962). *Group properties of differential equations.* Novosibirsk (in Russian).

Ovsiannikov, L. V. (1982). *Group analysis of differential equations.* Academic Press, New York.

Parker, D. F. (1980). The decay of saw-tooth solutions to the Burgers equation. *Proc. Roy. Soc.* **A369**, 409–24.

Peletier, L. A. (1970). Asymptotic behaviour of temperature profiles of a class of non-linear heat conduction problems. *Quart. J. Mech. Appl. Math.* **23**, 441–7.

Peletier, L. A. (1971). Asymptotic behaviour of solutions of the porous media equation. *SIAM J. Appl. Math.* **21**, 542–51.

Pelinovskii, E. N. and Fridman, V. E. (1974). Explosive instability in nonlinear waves in media with negative viscosity. *J. Appl. Math. and Mech.* (*PMM*) **38**, 940–4.

Penel, P. and Brauner, C. M. (1974). Identification of parameters in a nonlinear selfconsistent system including a Burgers equation. *J. Math. Anal. and Appl.* **45**, 654–81.

Protter, M. H. and Weinberger, H. F. (1967). *Maximum principles in differential equations.* Prentice Hall, Englewood Cliffs, N. J.

Qian, J. (1984). Numerical experiments on one-dimensional model of turbulence. *Phys. Fluids* **27**, 1957–65.

Richtmyer, R. D. and Morton, K. W. (1967). *Difference methods for initial value problems.* Interscience, New York.

Rodin, E. Y. (1969). A Riccati solution for Burgers equations. *Quart. Appl. Math.* **27**, 541–5.

Rodin , E. Y. (1970). On some approximate and exact solutions of boundary value problems for Burgers equation. *J. Math. Anal. Appl.* **30**, 401–14.

Rogers, C. (1983). Linked Bäcklund transformations and applications to nonlinear boundary value problems. University of Waterloo, Ontario, Dept. of Applied Mathematics (Preprint).

Rogers, C. and Ames, W. F. (1986). Nonlinear initial and boundary value problems in science and engineering. Academic Press (to be published).

Rogers, C. and Sachdev, P. L. (1984). The Burgers hierarchy: on nonlinear initial and boundary value problems. *Il Nuovo Cimento* **83B**, 127–34.

Rogers, C. and Shadwick, W. F. (1982). *Bäcklund transformations and their applications.* Academic Press, New York.

Rogers, C., Stallybrass, M. P. and Clements, D. L. (1983). On two-phase filtration under gravity and with boundary infiltration: application of a Bäcklund transformation. *Nonlinear Analysis, Theory, Methods and Applications* **7**, 785–99.

Romanova, N. N. (1970). The vertical propagation of short acoustic waves in the real atmosphere. *Atmos. and Oceanic Phys.* **6**, 73–8.

Rosenau, P. and Kamin, S. (1982). Nonlinear diffusion in a finite mass medium. *Comm. Pure and Appl. Math.* **35**, 113–27.

Rosenau, P. (1982). A nonlinear thermal wave in a reacting medium. *Physica* **5D**, 136–44.

Rosenbloom, P C. and Widder, D. V. (1959). Expansions in terms of heat polynomials and associated functions. *Trans. Amer. Math. Soc.* **92**, 220–66.

Rott, N. (1978). The description of simple waves by particle displacement. *J. Appl. Math. and Phys.* (*ZAMP*) **29**, 178–89.

Rudenko, O. V. and Soluyan, S. I. (1977). *Theoretical foundations of nonlinear acoustics* (*English translation by R. T. Beyer*). Consultants Bureau (Plenum Press), New York.

Sachdev, P. L. and Seebass, A. R. (1973). Propagation of spherical and cylindrical N-waves. *J. Fluid. Mech.* **58**, 197–205.

Sachdev, P. L. (1976a). A class of exact solutions of boundary value problems for Burgers equation. *Comp. and Math. with Appls.* **2**, 111–6.

Sachdev, P. L. (1976b). Some exact solutions of Burgers-type equations. *Quart. Appl. Math.* **34**, 118–22.

Sachdev, P. L. (1978). A generalised Cole-Hopf transformation for nonlinear parabolic and hyperbolic equations. *J. Appl. Math. and Phys.* (*ZAMP*) **29**, 963–70.

Sachdev, P. L. (1979). Self-similar solutions of generalised Burgers equation with damping. *Professor P. L. Bhatnagar Commemoration Volume.* National Academy of Sciences, India, 73–80.

Sachdev, P. L. and Nair, K. R. C. (1987). Generalised Burgers equations and Euler–Painlevé transcendents–II. *J. Math. Phys.* (to appear).

Sachdev, P. L., Nair, K. R. C. and Tikekar, V. G. (1986). Generalised Burgers equations and Euler Painlevé transcendents–I. *J. Math. Phys.* **27**, 1506–22.

Sachdev, P. L. and Reddy, A. V. (1982). Some exact solutions describing unsteady plane gas flows with shocks. *Quart. Appl. Math.* **40**, 249–72.

Sachdev, P. L., Tikekar, V. G. and Nair, K. R. C. (1986). Evolution and decay of spherical and cylindrical *N* waves. *J. Fluid Mech.* **172**, 347–71.

Schindler, G. M. (1970). Simple waves in multidimensional gas flow. *SIAM J. Appl. Math.* **19**, 390–407.

Scott, J. F. (1981a). Uniform asymptotics for spherical and cylindrical nonlinear acoustic waves generated by a sinusoidal source. *Proc. Roy. Soc.* **A375**, 211–30.

Scott, J. F. (1981b). The long time asymptotics of solutions to the generalised Burgers equation. *Proc. Roy. Soc.* **A373**, 443–56.

Sedov, L. I. (1946). Propagation of strong explosion waves. *Prikl. Math. Mekh.* **10**, 241–50.

Sedov, L. I. (1982). *Similarity and dimensional methods in mechanics.* Mir Publishers, Moscow.

Serrin, J. (1967). Asymptotic behaviour of velocity profiles in the Prandtl boundary layer theory. *Proc. Roy. Soc.* **A299**, 491–507.

Seshadri, V. S. and Sachdev, P. L. (1977). Quasi-simple wave solutions for acoustic gravity waves. *Phys. Fluids.* **20**, 888–94.

Shampine, L. F. (1973). Concentration dependent diffusion I. *Quart. Appl. Math.* **30**, 441–52.

Smith, G. D. (1978). *Numerical solution of partial differential equations.* Oxford University Press.

Smoller, J. (1983). *Shock waves and reaction–diffusion equations.* Springer-Verlag, New York.

Swan, G. W. (1977). Exact fundamental solutions of linear parabolic equations with spatially varying coefficients. *Bull. Math. Biol.* **39**, 435–51.

Taylor, G. I. (1910). The conditions necessary for discontinuous motion in gases. *Proc. Roy. Soc.* **A84**, 371–77.

Taylor, G. I. (1950). The formation of a blast wave by a very intense explosion. I. Theoretical discussion. *Proc. Roy. Soc.* **A201**, 159–74.

Titchmarsh, E. C. (1962). *Eigen-function expansions associated with second-order differential equations*, Vols. I and II. Clarendon Press, Oxford.

Ton, B. A. (1975). On the behaviour of the solution of the Burgers equation as the viscosity goes to zero. *J. Math. Anal. Appl.* **49**, 713–20.

Tychonov, A. N. and Samarski, A. A. (1964). *Partial Differential Equations of Mathematical Physics*, Vol. I. Holden-Day, London.

Van Duyn, C. J., and Peletier, L. A. (1977a). Asymptotic behaviour of solutions of a nonlinear diffusion equation. *Arch. Rat. Mech. and Anal.* **65**, 363–77.

Van Duyn, C. J. and Peletier, L. A. (1977b). A class of similarity solutions of the nonlinear diffusion equation. *Nonlinear Analysis, Theory, Methods and Applications* **1**, 223–33.

Van Dyke, M. D. (1975). *Perturbation methods in fluid mechanics.* Parabolic Press, Stanford.

Walsh, R. A. (1969). Initial value problems associated with $u_t(x,t) = \delta u_{xx}(x,t) - u(x,t)u_x(x,t)$. *J. Math. Anal. and Appl.* **26**, 235–47.

Whitham, G. B. (1952). The flow pattern of a supersonic projectile. *Comm. Pure Appl. Math.* **5**, 301–48.

Whitham, G. B. (1974). *Linear and nonlinear waves.* Wiley, New York.

Whitham, G. B. (1984). Comments on periodic waves and solitons. *IMA J. Appl. Math.* **32**, 353–66.

Widder, D. V. (1956). Integral transforms related to heat conduction. *Ann. Math.* **42**, 279–305.

Widder, D. V. (1962). Analytic solutions of the heat equation. *Duke Math. J.* **29**, 497–503.

Widder, D. V. (1975). *The heat equation.* Academic Press, New York.

Zakharov, N. S. and Korobeinikov, V. P. (1980). Group analysis of the generalised Korteweg–de Vries–Burgers equations. *J. Appl. Math. Mech. (PMM)* **44**, 668–71.

Zel'dovich, Ya. B. and Raizer, Yu. P. (1966). *Physics of shock waves and high temperature hydro-dynamic phenomena*, Vol. II. Academic Press, New York.

Author index

Subject index